NISTIR 7557

CALCULATIONS OF FLUENCE RATES OF UNSCATTERED X- AND GAMMA-RAY PHOTONS EMERGING FROM MODEL SPHERES OF SPECIAL NUCLEAR MATERIAL

Stephen Seltzer

National Institute of Standards and Technology
Gaithersburg, MD 20899

January 2009

U.S. DEPARTMENT OF COMMERCE
Carlos M. Gutierrez, Secretary

NATIONAL INSTITUTE OF STANDARDS AND TECHNOLOGY
Patrick Gallagher, Deputy Director

Calculations of Fluence Rates of Unscattered X- and Gamma-Ray Photons Emerging From Model Spheres of Special Nuclear Material

Stephen M. Seltzer
Ionizing Radiation Division
National Institute of Standards and Technology
Gaithersburg, MD 20899

ABSTRACT

Calculations have been done to develop estimates of the fluence rate of unscattered x and gamma rays emerging from two model spheres of special nuclear material: one of uranium enriched to 93.5 % (mass fraction) of ^{235}U; one of plutonium enriched to 6.0 % (mass fraction) of ^{240}Pu.

Key Words: fluence rate; photons; plutonium; radionuclides; spheres; uranium

Introduction

This report describes calculations to estimate the fluence rates of unscattered x and gamma rays emerging from spheres of special nuclear material. The results reported here are based on models with the following major parameters: (1) a 1 kg sphere of uranium originally enriched to 93.5 % (mass fraction) of ^{235}U and assumed to be 50 y old; and (2) a 0.5 kg sphere of plutonium originally enriched to 6 % (mass fraction) of ^{240}Pu and assumed to be up to 50 y old. The calculations focus on the fluence rate of the emergent unscattered x- and gamma-ray lines at a distance of 1 m from the center of the spheres, assumed to be surrounded by air. Thus the reported results pertain to the discrete line energies in the spectrum of emergent photons.

Method of Calculation

The assumed original compositions of the spheres are listed in Table 1. Note that the mass densities of actual spheres could be slightly lower (see, *e.g.*, Sapir *et al.* (1998)), perhaps by about 1 % or more. However, such a difference will have a negligible effect on the results calculated here, as the transport of the photons in such spheres to a point 1 m from the center is essentially governed only by the mass of the sphere. That is, the absorption and scattering depends on the product of attenuation coefficients and pathlengths, and this product (number of mean-free paths) is independent of the density. For the assumed spherically symmetric geometry, even a 3 % change in mass density results in a change of only 1 % in the radius of the sphere, and such a change compared to the 1 m radius of the scoring surface is expected to produce negligible effects.

Table 1. Assumed initial parameters for the spheres.

	U sphere	Pu sphere
Mass density	18.95 g/cm^3	19.84 g/cm^3
Radius	2.327 cm	1.819 cm
Composition (mass fraction):	0.7 % ^{234}U	0.05 % ^{238}Pu
	93.5 % ^{235}U	93.57 % ^{239}Pu
	5.8 % ^{238}U	6.00 % ^{240}Pu
		0.29 % ^{241}Pu
		0.02 % ^{242}Pu
		0.11 % ^{241}Am

The compositions given in Table 1 were adopted based on best estimates suggested by a number of unclassified documents (see, *e.g.*, Stefáka *et al.*, 2008; Valković, 2006; Nguyen and Zsigrai, 2006; Hollas *et al.*, 2005; Phillips *et al.*, 2005; Larsson and Haslip, 2004; Fetter et al., 1990).

The calculations were carried out in stages. First, decay data for the radionuclides involved were assembled from the on-line databases of the Brookhaven National Nuclear Data Center (NuDat 2.4, http://www.nndc.bnl.gov/nudat2/indx_dec.jsp, accessed March-April, 2008). In a very few cases of questionable, very minor, gamma-ray line energies and intensities, the Lawrence Berkeley National Laboratory on-line Table of Isotopes (http://ie.lbl.gov/education/isotopes.htm, accessed April 2008) was consulted to help make a decision.

The relevant assumed decay-chain data, including the symbol of the radionuclide, its half-life, and its branching fractions, are depicted in Figs. 1a to 1d. For decay calculations it was decided to ignore any branching whose fraction was less than 1 %, mainly to simplify the decay-chain activity calculations. We assume that little important relevant information has been sacrificed by this simplification; the neglected radionuclides are indicated with a dashed box in Fig. 1. Shown in the cutout in Fig. 1a is the branching decay of ^{234m}Pa (half life of 1.159 min), which decays by beta emission (99.84 %) to ^{234}U and by isomeric transition (0.16 %) to ^{234}Pa (half life 6.70 h) that then decays by beta emission (100 %) to ^{234}U. According to the adopted procedure, the decay of ^{234m}Pa to ^{234}U through the isomeric transition to ^{234}Pa has been ignored.

The data used in the calculations for the emission spectra of photons with energies above 1 keV for the radionuclides of interest are listed in Appendix A.

Uranium Series

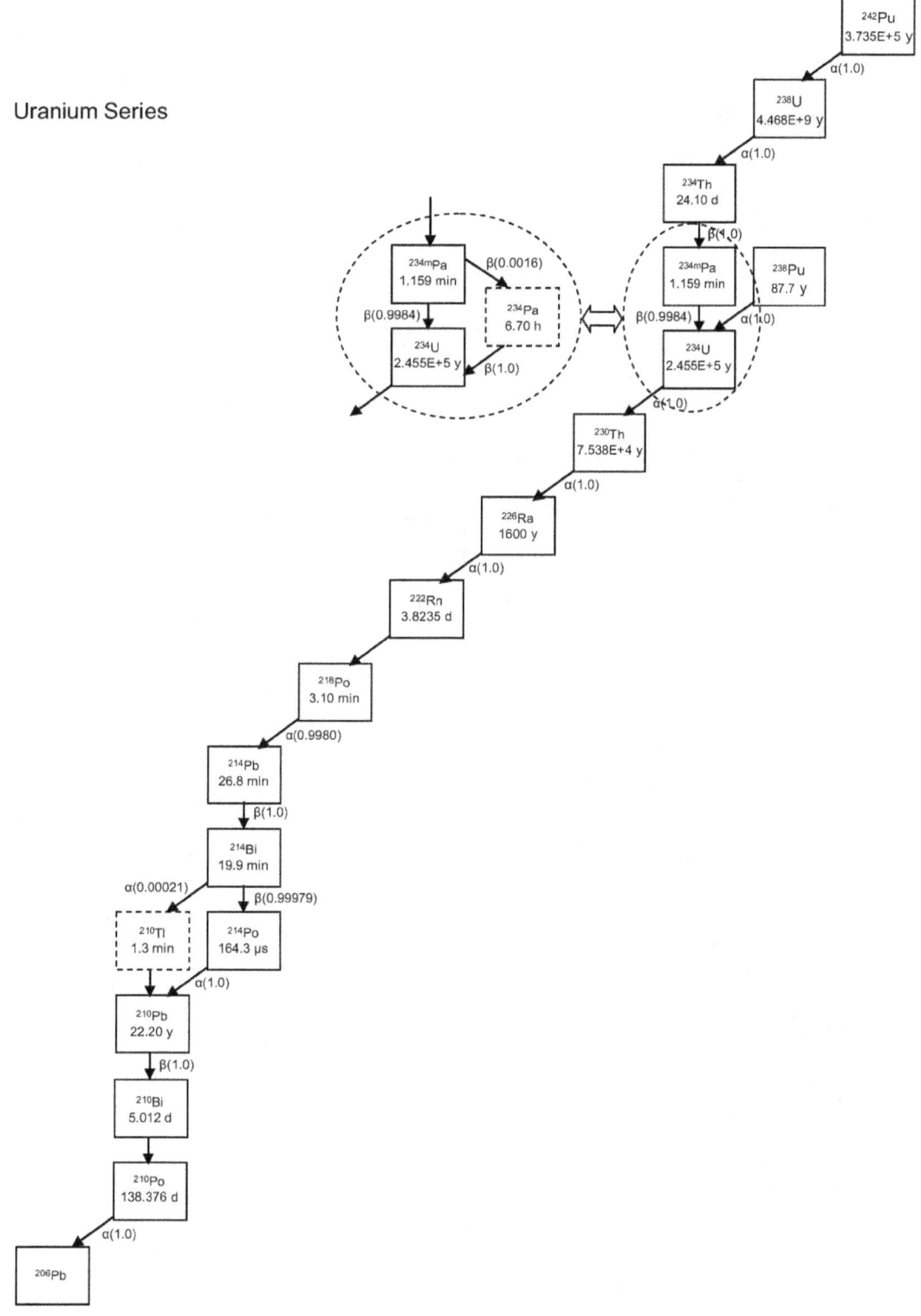

Fig. 1. Decay data giving assumed half-lives and branching fractions. The dotted squares indicate a branch that was ignored in the calculations.
a. Uranium series

3

Neptunium Series

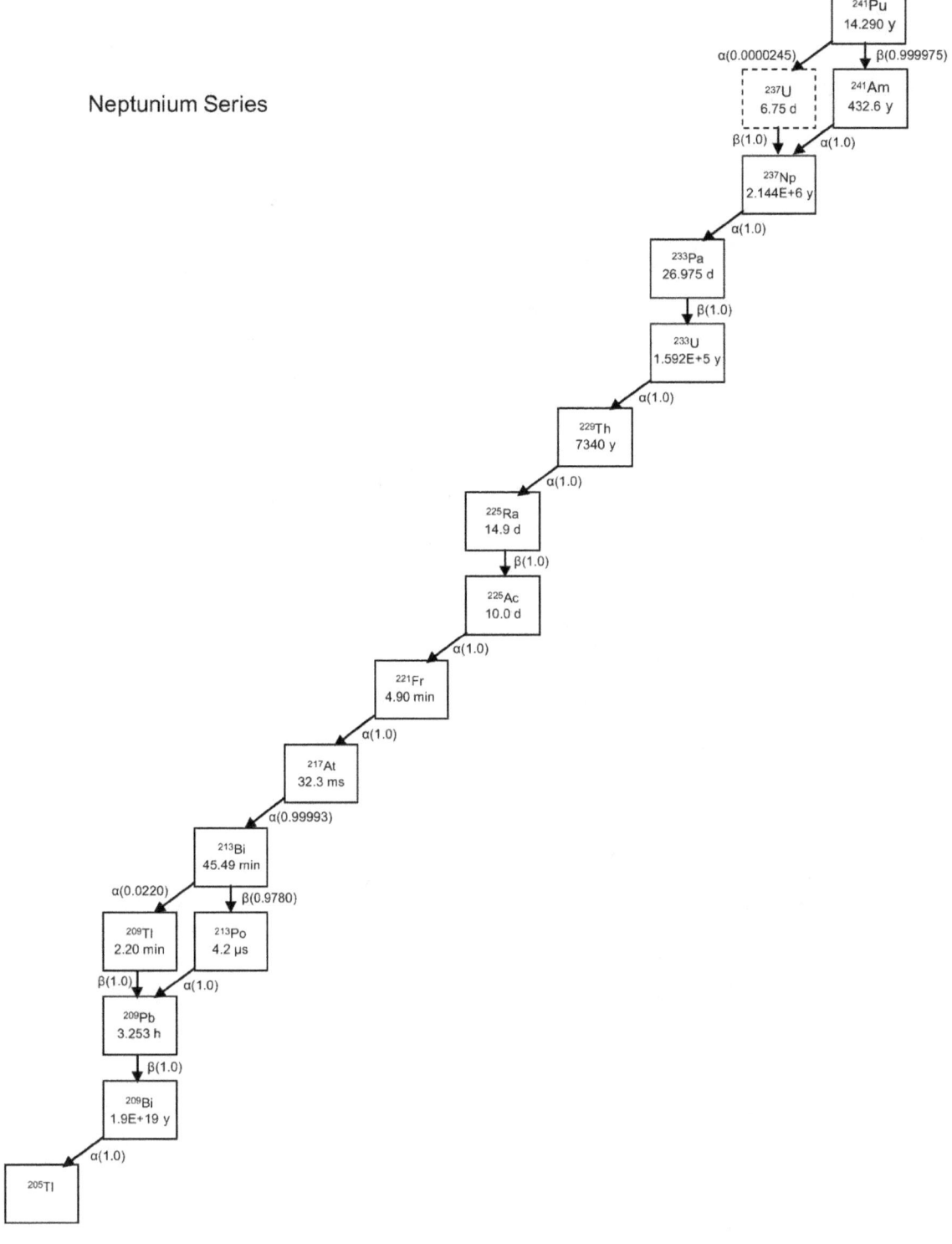

Fig. 1b. Neptunium series.

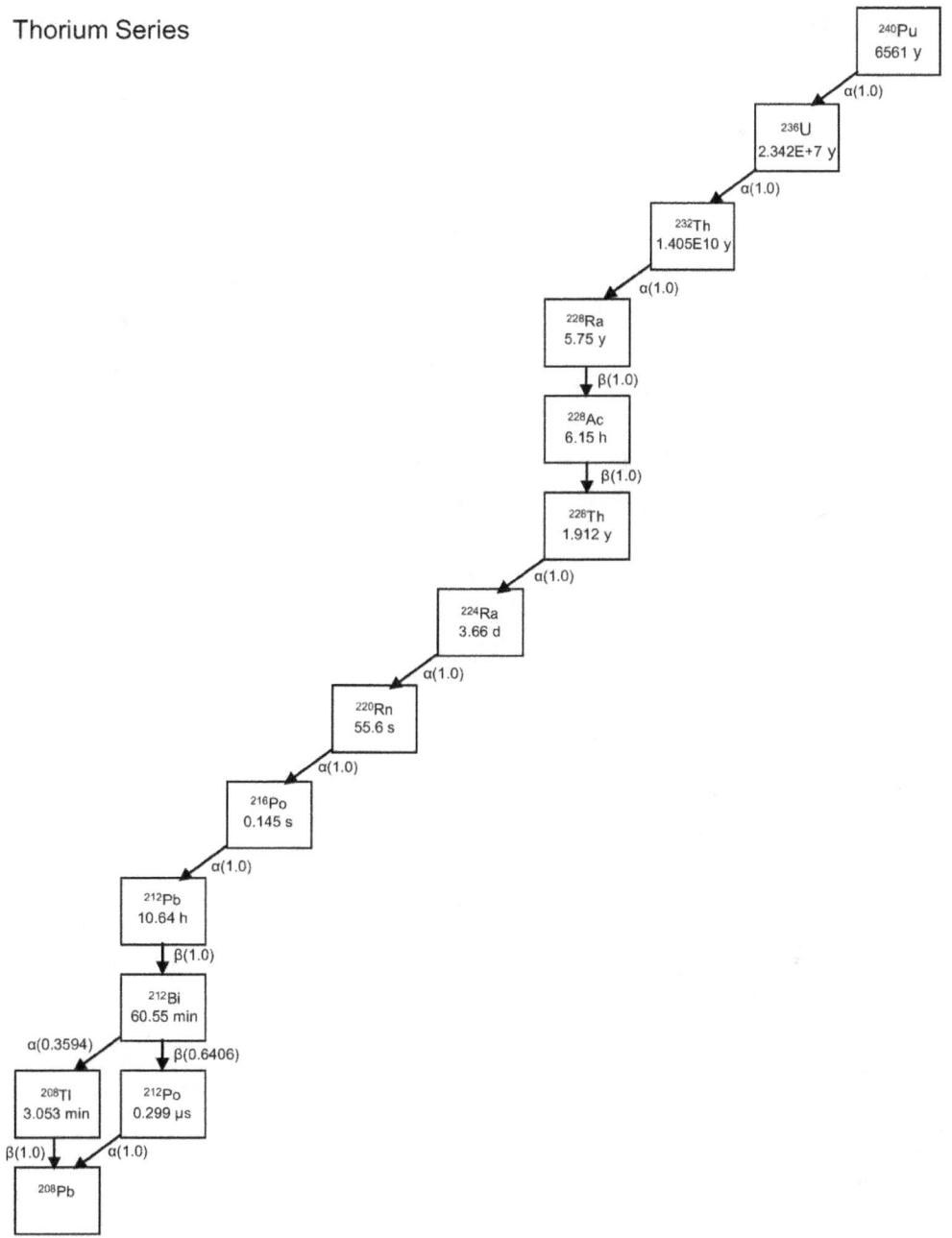

Thorium Series

^{240}Pu
6561 y

α(1.0)

^{236}U
2.342E+7 y

α(1.0)

^{232}Th
1.405E10 y

α(1.0)

^{228}Ra
5.75 y

β(1.0)

^{228}Ac
6.15 h

β(1.0)

^{228}Th
1.912 y

α(1.0)

^{224}Ra
3.66 d

α(1.0)

^{220}Rn
55.6 s

α(1.0)

^{216}Po
0.145 s

α(1.0)

^{212}Pb
10.64 h

β(1.0)

^{212}Bi
60.55 min

α(0.3594)   β(0.6406)

^{208}Tl
3.053 min

^{212}Po
0.299 μs

β(1.0)   α(1.0)

^{208}Pb

Fig. 1c. Thorium series.

5

Actinium Series

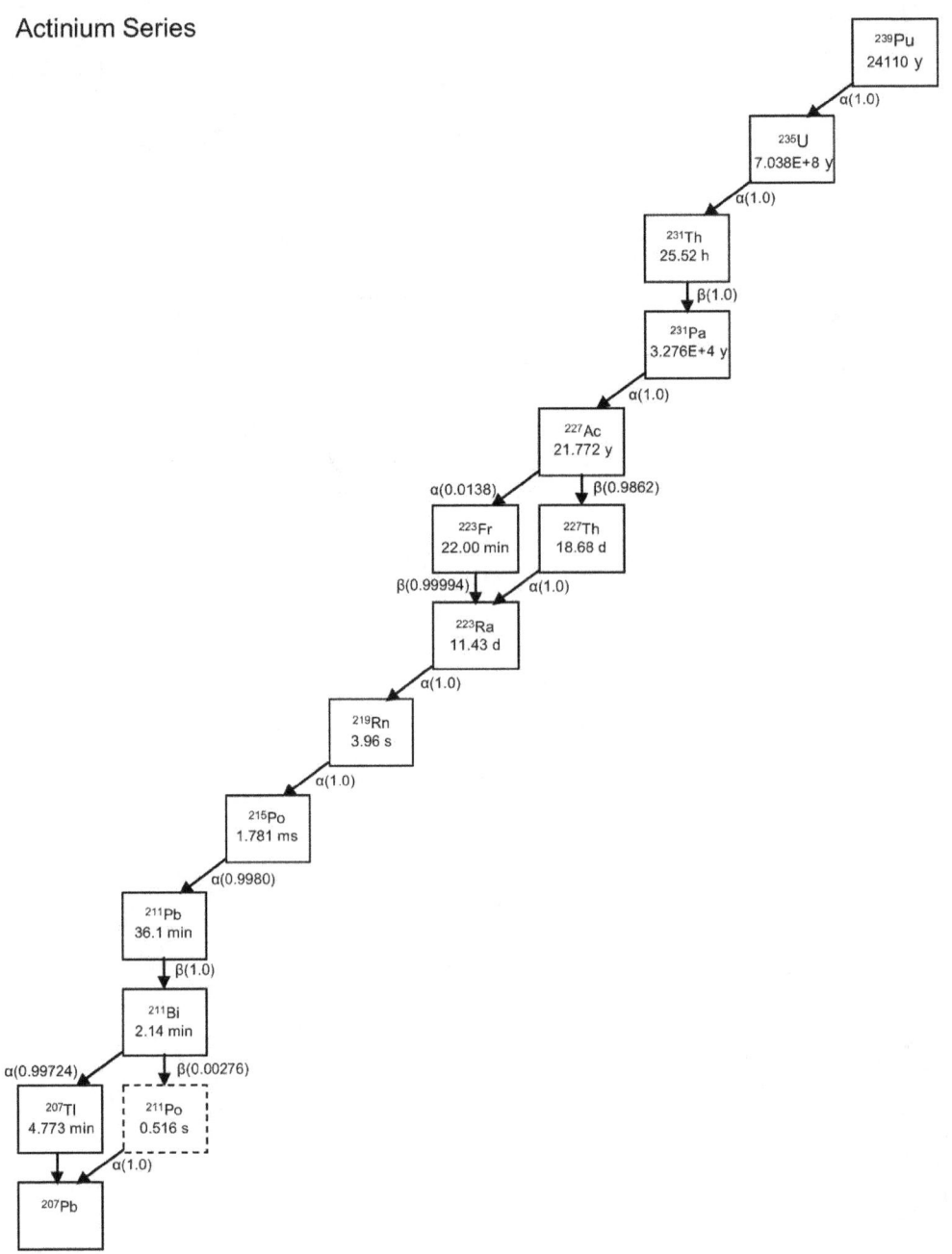

Fig. 1d. Actinium series.

Calculations of the activities of the radionuclides as a function of time were done by means of the usual Bateman equation (Bateman, 1910) for each original constituent radionuclide, assuming the concentrations of all progeny are initially zero. Thus the number of atoms n at time t from an initial number of the parent atoms n_1 is

$$n_i(t) = \lambda_1 \lambda_2 \cdots \lambda_{i-1} n_1(0) \sum_{j=1}^{i} \frac{\exp(-\lambda_j t)}{\prod_{\substack{k=1 \\ k \neq j}}^{i} (\lambda_k - \lambda_j)}, \tag{1}$$

where λ_i is the decay constant for the pertinent radionuclide. The relative atomic weights of the pertinent radionuclides were taken from Coursey *et al.* (2005). Note that the possible production of radionuclides through emission and absorption of neutrons is not included, and is assumed to be negligible for this problem. Complex decay (*i.e.*, multiple branching) occurs in all four decay series, but at most only once with our procedures. The associated branching fractions were inserted to multiply the λ terms in Eq. 1 just to the right of the equal sign that multiply $n_1(0)$. In the cases of complex decay, the chain was calculated for both branches, and the results for radionuclides below the split were then summed.

The activities at time t are given by $\lambda_i n_i(t)$, and can be evaluated for any t. The results for our spheres at $t = 10$ y, 25 y, and 50 y are given in Table 2.

Table 2. Activities of radionuclides in spheres at a function of age, from initial isotopic compositions given in Table 1.

a. Uranium

Nuclide	Activity $\lambda_i n_i$ / s^{-1}		
	10 y	25 y	50 y
U-238	7.213E+05	7.213E+05	7.213E+05
U-235	7.476E+07	7.476E+07	7.476E+07
Th-234	7.213E+05	7.213E+05	7.213E+05
Pa-234m	7.202E+05	7.202E+05	7.202E+05
U-234	1.611E+09	1.611E+09	1.611E+09
Th-231	7.476E+07	7.476E+07	7.476E+07
Pa-231	1.581E+04	3.953E+04	7.905E+04
Th-230	1.482E+05	3.704E+05	7.407E+05
Ac-227	2.269E+03	1.227E+04	3.950E+04
Th-227	2.206E+03	1.204E+04	3.886E+04
Ra-226	3.205E+02	1.999E+03	7.965E+03
Fr-223	3.131E+01	1.693E+02	5.451E+02
Ra-223	2.218E+03	1.217E+04	3.935E+04
Rn-222	3.195E+02	1.996E+03	7.961E+03
Rn-219	2.218E+03	1.217E+04	3.935E+04
Po-218	3.195E+02	1.996E+03	7.959E+03
Po-215	2.218E+03	1.217E+04	3.935E+04
Pb-214	3.195E+02	1.996E+03	7.959E+03
Bi-214	3.194E+02	1.995E+03	7.957E+03
Po-214	3.194E+02	1.995E+03	7.957E+03
Pb-211	2.218E+03	1.217E+04	3.935E+04
Bi-211	2.212E+03	1.214E+04	3.924E+04
Pb-210	3.077E+01	4.320E+02	2.926E+03
Bi-210	3.059E+01	4.310E+02	2.923E+03
Po-210	2.617E+01	4.054E+02	2.839E+03
Tl-207	2.212E+03	1.214E+04	3.924E+04

Table 2. Activities of radionuclides in spheres at a function of age, from initial isotopic compositions given in Table 1.

b. Plutonium

Nuclide	Activity $\lambda_i n_i$ / s^{-1}			Nuclide	Activity $\lambda_i n_i$ / s^{-1}		
	10 y	25 y	50 y		10 y	25 y	50 y
Pu-242	1.463E+07	1.463E+07	1.463E+07	Fr-223	1.514E-03	2.120E-02	1.435E-01
Pu-241	3.428E+12	1.656E+12	4.925E+11	Ra-223	1.060E-01	1.516E+00	1.034E+01
Am-241	1.387E+11	1.932E+11	2.231E+11	Rn-222	5.790E-02	8.798E-01	6.700E+00
Pu-240	2.517E+11	2.513E+11	2.506E+11	Fr-221	1.861E-03	3.814E-02	3.839E-01
Pu-239	1.073E+12	1.073E+12	1.072E+12	Rn-220	2.922E-06	4.786E-05	2.923E-04
U-238	2.270E-02	5.674E-02	1.135E-01	Rn-219	1.060E-01	1.516E+00	1.034E+01
Pu-238	2.927E+10	2.600E+10	2.134E+10	Po-218	5.789E-02	8.796E-01	6.699E+00
Np-237	3.465E+05	1.169E+06	2.883E+06	At-217	1.861E-03	3.814E-02	3.839E-01
U-236	7.453E+04	1.862E+05	3.719E+05	Po-216	2.922E-06	4.786E-05	2.923E-04
U-235	1.057E+04	2.643E+04	5.284E+04	Po-215	1.060E-01	1.516E+00	1.034E+01
Th-234	2.248E-02	5.653E-02	1.133E-01	Pb-214	5.789E-02	8.796E-01	6.699E+00
Pa-234m	2.245E-02	5.644E-02	1.131E-01	Bi-214	5.788E-02	8.794E-01	6.697E+00
U-234	8.600E+05	2.029E+06	3.694E+06	Po-214	5.788E-02	8.794E-01	6.697E+00
Pa-233	3.418E+05	1.162E+06	2.875E+06	Bi-213	1.861E-03	3.814E-02	3.839E-01
U-233	6.580E+00	5.426E+01	2.718E+02	Po-213	1.820E-03	3.730E-02	3.754E-01
Th-232	1.839E-05	1.149E-04	4.590E-04	Pb-212	2.921E-06	4.785E-05	2.923E-04
Th-231	1.057E+04	2.642E+04	5.283E+04	Bi-212	2.920E-06	4.785E-05	2.923E-04
Pa-231	1.118E+00	6.987E+00	2.794E+01	Po-212	1.871E-06	3.065E-05	1.872E-04
Th-230	4.006E+01	2.409E+02	9.049E+02	Pb-211	1.060E-01	1.516E+00	1.034E+01
Th-229	1.921E-03	3.864E-02	3.864E-01	Bi-211	1.057E-01	1.512E+00	1.031E+01
Ra-228	5.607E-06	6.270E-05	3.320E-04	Pb-210	4.261E-03	1.492E-01	2.003E+00
Ac-228	5.606E-06	6.269E-05	3.320E-04	Bi-210	4.228E-03	1.488E-01	2.000E+00
Th-228	2.936E-06	4.793E-05	2.925E-04	Po-210	3.449E-03	1.371E-01	1.922E+00
Ac-227	1.097E-01	1.536E+00	1.040E+01	Tl-209	4.094E-05	8.390E-04	8.445E-03
Th-227	1.059E-01	1.503E+00	1.022E+01	Pb-209	1.861E-03	3.814E-02	3.838E-01
Ra-226	5.817E-02	8.813E-01	6.706E+00	Bi-209	1.390E-21	9.161E-21	1.629E-19
Ra-225	1.885E-03	3.834E-02	3.849E-01	Tl-208	1.050E-06	1.720E-05	1.050E-04
Ac-225	1.861E-03	3.814E-02	3.839E-01	Tl-207	1.057E-01	1.512E+00	1.031E+01
Ra-224	2.922E-06	4.786E-05	2.923E-04				

The transport of photons was calculated with the Monte Carlo code ACCEPT from the Integrated Tiger Series Version 3 (Halbleib *et al.*, 1992). The code was altered to simulate an emission source uniformly distributed in the U or Pu sphere volume and to include unscattered photons in the fluence scores. The spheres were assumed to be centered in a sphere of dry air of density 1.197×10^{-3} g/cm^3 (*i.e.*, 20 °C, 101.325 kPa), and the fluence spectra were scored at a radius of 1 m in an outer air shell 1 mm thick. Our knowledge of the transport properties of photons (*e.g.*, attenuation coefficients) and electrons (*e.g.*, stopping powers) in U and in Pu does not distinguish among isotopes, so atomic cross sections were used (Berger and Hubbell, 1987). With up to 50 y of decay, the progenies with atomic numbers different from U or Pu are at sufficiently low levels that they can be assumed not to alter the photon-transport properties of the spheres.

Photon-source emission spectra can be obtained from the product of the calculated activities in Table 2 and the emission probabilities in Appendix A. However, direct simulations of the emission spectra in a Monte Carlo calculation would result in under-sampling of low-intensity lines and very poor scoring statistics for any low-energy photons, *i.e.*, the probability of emergence from the surface of the heavy-metal sphere becomes vanishingly small for photons emitted more than some few mean-free paths from the surface. Instead, the Monte Carlo calculations were done for uniformly distributed, isotropic emission of monoenergetic photons, with energies from 3.5 MeV for U and from 4.0 MeV for Pu, down to 30 keV for both spheres. One can go to lower energies, but the emergence probability becomes very small, and it was presumed that energies below those of the L-shell absorption-edge energies of U and Pu would be of little interest. For emitted photon energies of 500 keV and below, the source sampling was restricted to a volume of a spherical shell extending from about 7 mean-free paths below the sphere surface out to the sphere surface, and the results normalized to the total volume assuming that photons emitted at deeper source points have negligible probabilities of emergence.

Because it is the fluence rate of the emergent photon lines that would be useful for detection and identification of such nuclear material, this round of Monte Carlo calculations included only photon transport and ignored the bremsstrahlung resulting from beta emission and produced by secondary electrons set in motion by the source photons. These components are estimated to make a small contribution to the continuum. However, the continuum produced by source photons scattered in the spheres was scored, although not reported here. The advantage of this approach based on monoenergetic photon transport is that uncertainties can be controlled, and results for any source spectrum (but not sphere size) can be readily evaluated; its disadvantage lies in the neglect of presumably relatively small bremsstrahlung contributions to the continuum and to the production of characteristic x rays..

The line fluences at 1 m in air were based on 10^6 histories for each source energy, which resulted in estimated relative standard deviations of the mean of 1 % or less. The results are given in Table 3 and in Fig. 2. Note that for each monoenergetic source energy, there are associated secondary-radiation line fluences of annihilation quanta (produced following pair production) and of characteristic x rays (produced in photoelectric absorption).

Table 3. Fluence from photons of energy E at 1 m in air for the U and Pu spheres. Given are the fluence of the unscattered emergent source photons, $\Phi_o(E)$, the associated fluence of emergent annihilation quanta, $\Phi_{mc2}(0.511 \text{ MeV})$, and the fluences of emergent K-shell characteristic x rays, $\Phi_{K\beta1}(E_{K\beta1})$, $\Phi_{K\alpha1}(E_{K\alpha1})$, and $\Phi_{K\alpha2}(E_{K\alpha2})$. The fluences are given in units of cm^{-2}, normalized to one source photon emitted uniformly and isotropically in the sphere.

a. U sphere. The energies of the x-ray lines are: $E_{K\beta1}$=0.11130 MeV, $E_{K\alpha1}$=0.09844 MeV, and $E_{K\alpha2}$=0.09466 MeV.

E/MeV	Φ_o/cm^{-2}	Φ_{mc2}/cm^{-2}	$\Phi_{K\beta1}$/cm^{-2}	$\Phi_{K\alpha1}$/cm^{-2}	$\Phi_{K\alpha2}$/cm^{-2}
3.5	2.71E-06	2.87E-07	5.11E-09	8.02E-09	4.34E-09
3.0	2.69E-06	2.35E-07	4.92E-09	7.62E-09	4.19E-09
2.5	2.64E-06	1.75E-07	4.79E-09	7.17E-09	4.02E-09
2.0	2.52E-06	1.05E-07	4.67E-09	7.03E-09	3.89E-09
1.5	2.27E-06	3.18E-08	4.76E-09	7.26E-09	4.02E-09
1.0	1.70E-06	0	5.60E-09	8.69E-09	4.81E-09
0.8	1.36E-06	0	6.30E-09	9.67E-09	5.25E-09
0.6	9.50E-07	0	7.10E-09	1.09E-08	6.12E-09
0.5	7.20E-07	0	7.88E-09	1.19E-08	6.69E-09
0.4	4.92E-07	0	8.33E-09	1.28E-08	7.25E-09
0.3	2.75E-07	0	9.55E-09	1.44E-08	8.17E-09
0.2	1.09E-07	0	1.12E-08	1.69E-08	9.51E-09
0.15	5.39E-08	0	1.21E-08	1.85E-08	1.04E-08
0.116	2.83E-08	0	1.22E-08	1.95E-08	1.10E-08
0.115	1.10E-07	0	0	0	0
0.1	7.81E-08	0	0	0	0
0.08	5.70E-09	0	0	0	0
0.06	1.35E-09	0	0	0	0
0.05	5.25E-10	0	0	0	0
0.04	1.68E-10	0	0	0	0
0.03	3.78E-11	0	0	0	0

Table 3. Fluence from photons of energy E at 1 m in air for the U and Pu spheres. Given are the fluence of the unscattered emergent source photons, $\Phi_o(E)$, the associated fluence of emergent annihilation quanta, $\Phi_{mc2}(0.511\ MeV)$, and the fluences of emergent K-shell characteristic x rays, $\Phi_{K\beta1}(E_{K\beta1})$, $\Phi_{K\alpha1}(E_{K\alpha1})$, and $\Phi_{K\alpha2}(E_{K\alpha2})$. The fluences are given in units of cm^{-2}, normalized to one source photon emitted uniformly and isotropically in the sphere.

b. Pu sphere. The energies of the x-ray lines are: $E_{K\beta1}$=0.11723 MeV, $E_{K\alpha1}$=0.1037 MeV, and $E_{K\alpha2}$=0.09953 MeV.

E/MeV	Φ_o/cm^{-2}	Φ_{mc2}/cm^{-2}	$\Phi_{K\beta1}$/cm^{-2}	$\Phi_{K\alpha1}$/cm^{-2}	$\Phi_{K\alpha2}$/cm^{-2}
4.0	3.08E-06	3.60E-07	6.53E-09	9.79E-09	5.51E-09
3.5	3.08E-06	3.11E-07	6.25E-09	9.49E-09	5.20E-09
3.0	3.05E-06	2.56E-07	6.02E-09	8.84E-09	5.10E-09
2.5	3.00E-06	1.92E-07	5.76E-09	8.77E-09	4.80E-09
2.0	2.87E-06	1.16E-07	5.77E-09	8.36E-09	4.85E-09
1.5	2.59E-06	3.61E-08	5.84E-09	8.76E-09	4.94E-09
1.0	1.95E-06	0	6.95E-09	1.05E-08	5.94E-09
0.8	1.57E-06	0	8.00E-09	1.19E-08	6.76E-09
0.6	1.09E-06	0	9.03E-09	1.39E-08	7.76E-09
0.5	8.23E-07	0	9.99E-09	1.49E-08	8.49E-09
0.4	5.61E-07	0	1.08E-08	1.64E-08	9.23E-09
0.3	3.13E-07	0	1.24E-08	1.84E-08	1.04E-08
0.2	1.24E-07	0	1.44E-08	2.15E-08	1.21E-08
0.15	6.15E-08	0	1.54E-08	2.37E-08	1.33E-08
0.122	3.69E-08	0	1.55E-08	2.45E-08	1.38E-08
0.121	1.40E-07	0	0	0	0
0.1	8.71E-08	0	0	0	0
0.08	7.14E-09	0	0	0	0
0.06	1.69E-09	0	0	0	0
0.05	6.59E-10	0	0	0	0
0.04	2.13E-10	0	0	0	0
0.03	4.86E-11	0	0	0	0

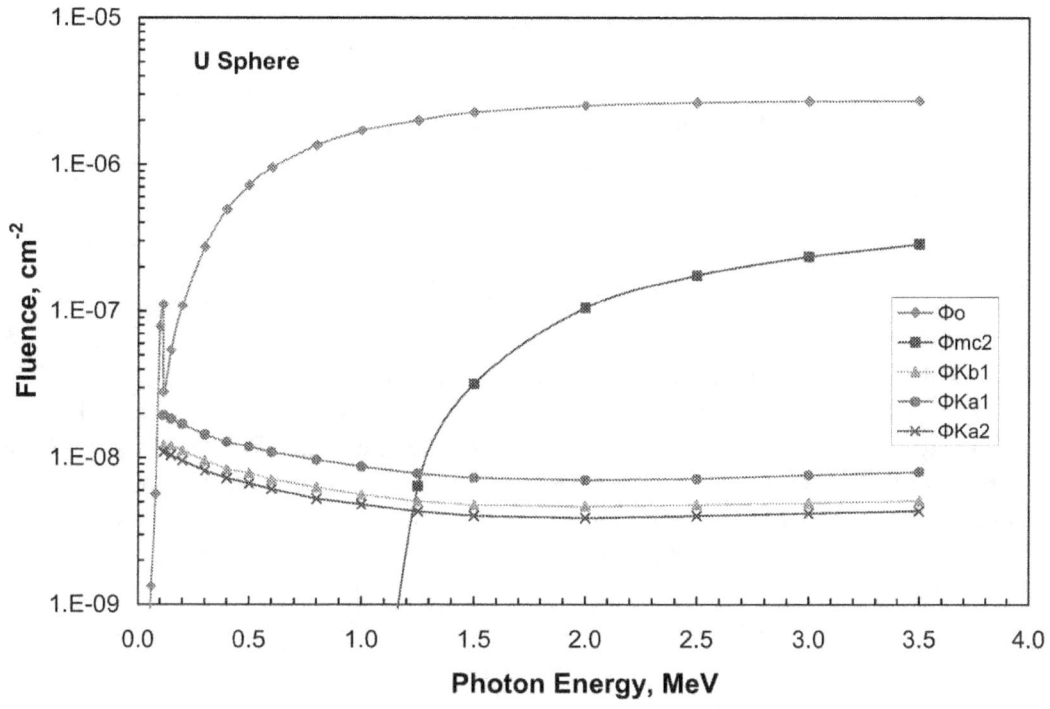

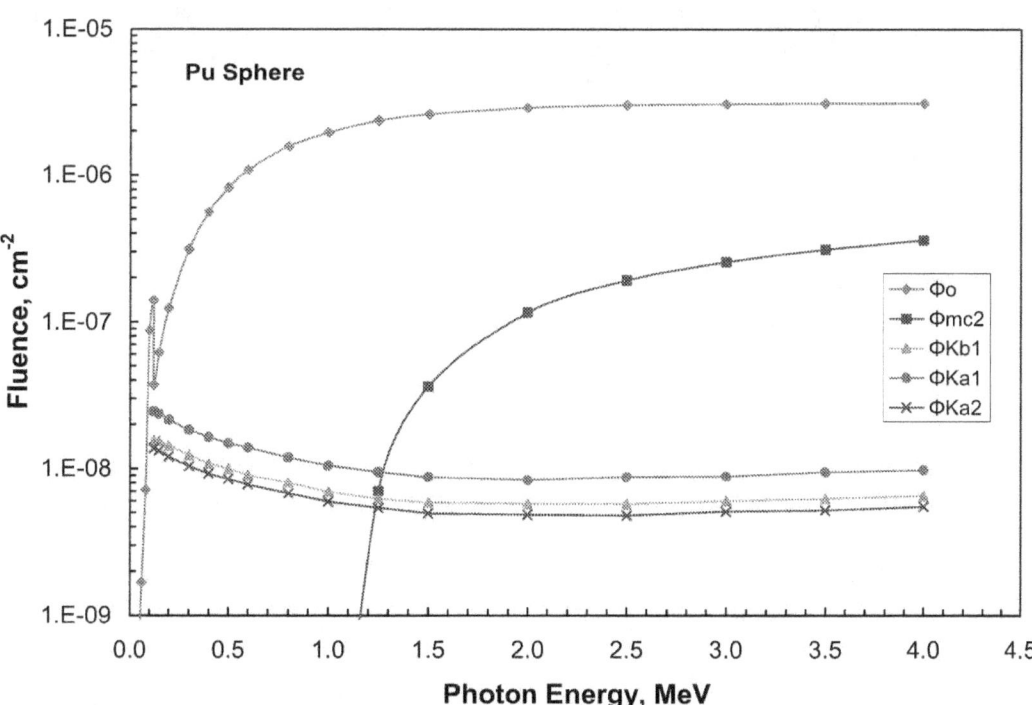

Fig. 2. Fluence from monoenergetic photons at 1 m in air for the U and Pu spheres, normalized to one internally emitted source photon (see Tables 3a and 3b).

Results

The final stage is to combine our results as the product of the calculated activities in Table 2, the inherent emission probabilities in Appendix A, and interpolated values of the unscattered emergent probabilites in Table 3. Because of the large number of photons emitted by the radionuclides involved, even considering only those with energies above 30 keV leads to 1182 lines from the U sphere and 2490 lines from the Pu sphere. Complete results are available from the author in the form of lists of energy-fluence-rate pairs for each assumed sphere and age, ordered either in increasing photon energy or in decreasing fluence rate.

In an effort to communicate the scale of the results, Figs. 3 and 4 plot the unscattered fluence rates for 50-year-old spheres, simply as points, over about 6 orders of magnitude. Note that these plots are not estimates of detector pulse-height distributions, because the effects of background, detector efficiency and resolution, and counting statistics are not included. Rather the points simply represent the fluence rates of the discrete line energies. The plots indicate the level of fluence rates obtained, and perhaps identify a few of the more-prominent lines. However, any interpretation in terms of detectability should take into account the possible overlap of lines close in energy, due to the inherent spreading of these lines by the detector resolution and response function, and the influences of counting statistics, contributions from source photons scattered in the spheres and from emergent bremsstrahlung, and contributions of environmental sources.

The higher fluence rates are listed in Table 4, in order of descending fluence rate. There are negligible differences in these fluence rates for the ^{235}U (93.5 %) sphere of ages 10 y, 25 y, and 50 y (Table 4a). However, the decay dynamics of the ^{240}Pu (6.0 %) sphere show a bit more complexity, and separate tables are given for ages of 50 y (Table 4b), 25 y (Table 4c), and 10 y (Table 4d).

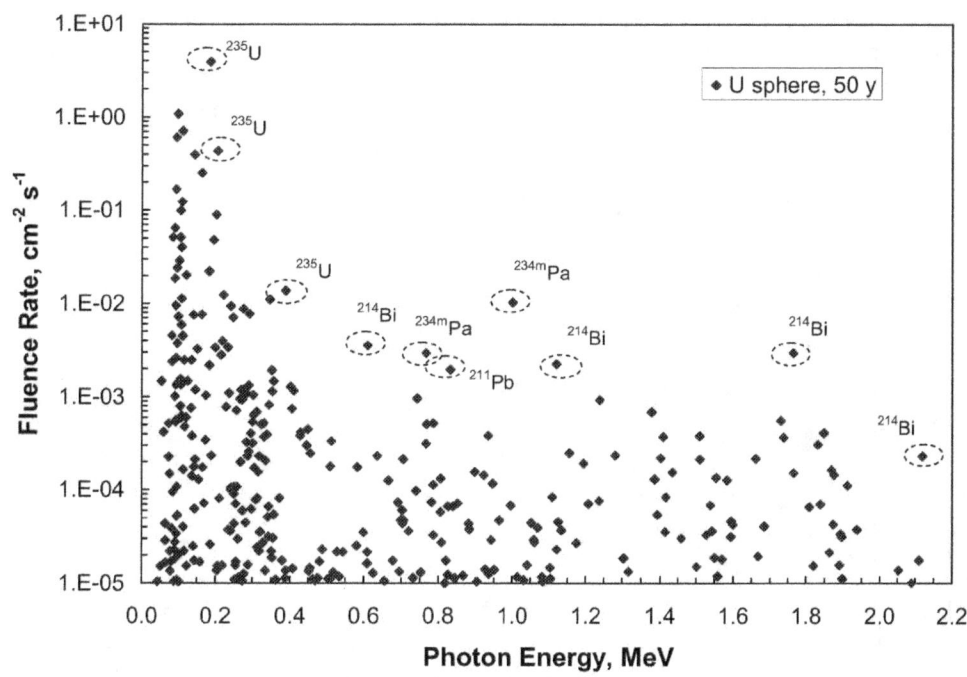

Fig. 3. Unscattered fluence rate at 1 m in air of photon lines emergent from the ^{235}U (93.5 %) sphere, age 50 y, of Table 1.

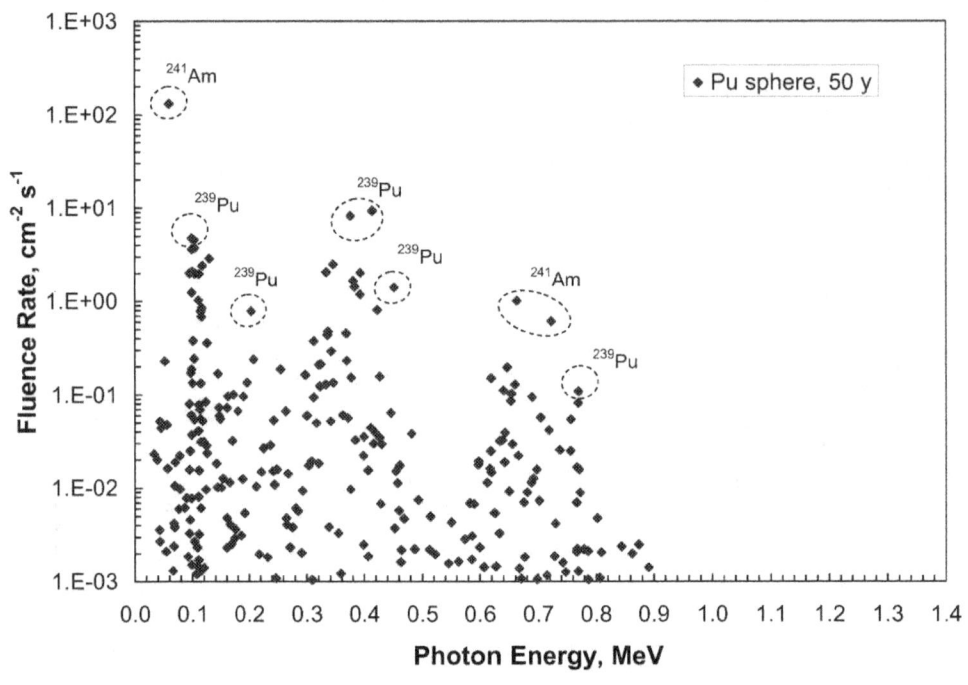

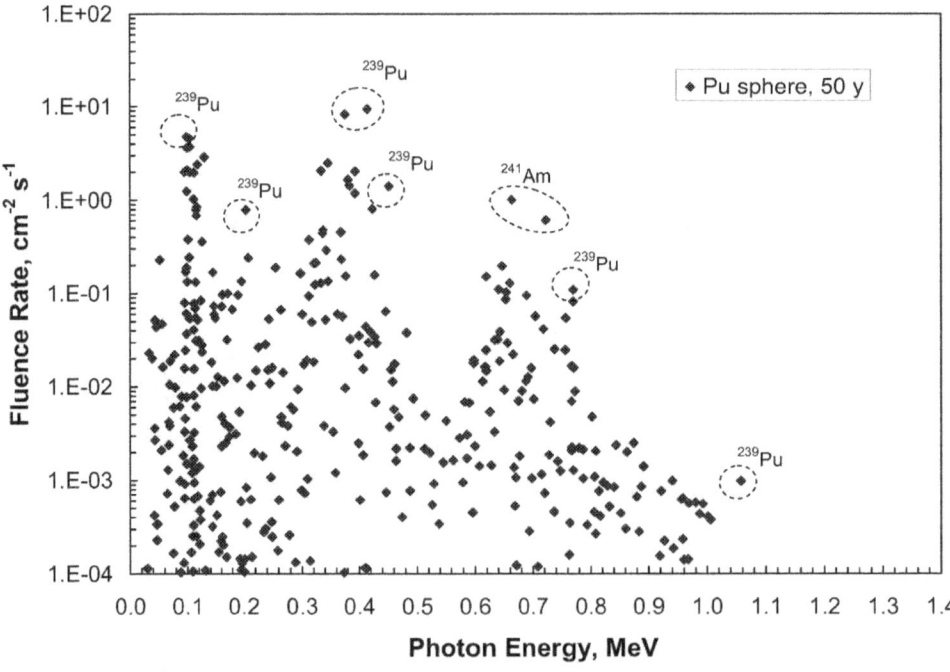

Fig. 4. Unscattered fluence rate at 1 m in air of photon lines emergent from the [240]Pu (6.0 %) sphere, age 50 y, of Table 1. Both plots are of the same data but with different ordinate scales.

Table 4. Fluence rate from unscattered photons of energy E at 1 m in air for the U and Pu spheres.

a. ^{235}U (93.5 %) 1 kg sphere; fluence rates greater than 10^{-2} cm^{-2} s^{-1}. Except for changes less than 0.1 % in the fluence rates of the U characteristic x rays, these results hold for sphere ages of 10 y, 25 y, and 50 y, and the total fluence rates of all 1182 lines calculated are about 8.5 cm^{-2} s^{-1} for these ages.

E/MeV	$\dot{\Phi}$ /cm^{-2} s^{-1}	Nuclide
1.857E-01	3.899E+00	U235
9.844E-02	1.084E+00	U Kα1
1.113E-01	7.151E-01	U Kβ1
9.466E-02	6.097E-01	U Kα2
2.053E-01	4.346E-01	U235
1.438E-01	3.973E-01	U235
1.633E-01	2.528E-01	U235
9.335E-02	1.671E-01	U235
1.092E-01	1.234E-01	U235
1.056E-01	9.947E-02	U235
2.021E-01	9.025E-02	U235
8.996E-02	6.486E-02	U235
8.421E-02	5.142E-02	TH231
1.048E-01	5.109E-02	U235
1.949E-01	4.828E-02	U235
1.086E-01	4.028E-02	U235
1.023E-01	2.910E-02	TH231
9.586E-02	2.413E-02	TH231
1.826E-01	2.225E-02	U235
1.209E-01	2.024E-02	U234
8.995E-02	1.867E-02	TH231
3.878E-01	1.389E-02	U235
2.214E-01	1.244E-02	U235
1.084E-01	1.136E-02	TH231
3.459E-01	1.109E-02	U235
1.001E+00	1.032E-02	PA234m

Table 4. Fluence rate from unscattered photons of energy *E* at 1 m in air for the U and Pu spheres.

b. ^{240}Pu (6.0 %) 0.5 kg sphere of age 50 y; fluence rates greater than 10^{-1} cm^{-2} s^{-1}. Total fluence rate for all 2490 lines calculated is 211 cm^{-2} s^{-1}.

E/MeV	$\dot{\Phi}$ /cm^{-2} s^{-1}	Nuclide	*E*/MeV	$\dot{\Phi}$ /cm^{-2} s^{-1}	Nuclide
5.954E-02	1.311E+02	AM241	3.354E-01	4.388E-01	AM241
4.137E-01	9.378E+00	PU239	1.011E-01	3.792E-01	AM241
3.751E-01	8.272E+00	PU239	3.119E-01	3.767E-01	PA233
9.843E-02	4.765E+00	PU239	1.253E-01	3.589E-01	AM241
1.030E-01	4.576E+00	AM241	3.415E-01	2.921E-01	PU239
1.037E-01	3.736E+00	Pu Kα1	1.031E-01	2.446E-01	PU239
9.897E-02	3.634E+00	AM241	2.080E-01	2.403E-01	AM241
1.293E-01	2.882E+00	PU239	3.687E-01	2.324E-01	AM241
3.450E-01	2.505E+00	PU239	5.162E-02	2.293E-01	PU239
1.172E-01	2.423E+00	Pu Kβ1	3.238E-01	2.130E-01	PU239
9.953E-02	2.102E+00	Pu Kα2	3.209E-01	2.100E-01	PU239
3.328E-01	2.067E+00	PU239	6.459E-01	1.970E-01	PU239
3.931E-01	2.038E+00	PU239	9.843E-02	1.883E-01	PU241A
9.465E-02	2.007E+00	PU239	2.554E-01	1.880E-01	PU239
1.042E-01	1.999E+00	PU240	9.707E-02	1.715E-01	AM241
1.113E-01	1.980E+00	PU239	1.442E-01	1.693E-01	PU239
3.802E-01	1.667E+00	PU239	2.975E-01	1.641E-01	PU239
3.828E-01	1.434E+00	PU239	4.267E-01	1.575E-01	PU239
4.515E-01	1.412E+00	PU239	3.767E-01	1.541E-01	AM241
9.878E-02	1.244E+00	PU239	6.190E-01	1.510E-01	AM241
3.925E-01	1.190E+00	PU239	3.450E-01	1.352E-01	PU239
1.104E-01	1.028E+00	PU239	1.957E-01	1.350E-01	PU239
6.624E-01	1.015E+00	AM241	9.985E-02	1.340E-01	PU238
1.163E-01	8.483E-01	PU239	1.142E-01	1.328E-01	AM241
4.226E-01	8.107E-01	PU239	3.324E-01	1.294E-01	AM241
2.036E-01	7.889E-01	PU239	6.589E-01	1.291E-01	PU239
1.144E-01	7.873E-01	PU239	3.225E-01	1.239E-01	AM241
1.154E-01	6.863E-01	PU239	6.400E-01	1.113E-01	PU239
7.220E-01	6.100E-01	AM241	7.694E-01	1.096E-01	PU239
3.361E-01	4.783E-01	PU239	6.530E-01	1.032E-01	AM241
3.671E-01	4.542E-01	PU239	1.714E-01	1.007E-01	PU239
3.686E-01	4.527E-01	PU239			

Table 4. Fluence rate from unscattered photons of energy E at 1 m in air for the U and Pu spheres.

c. ^{240}Pu (6.0 %) 0.5 kg sphere of age 25 y; fluence rates greater than 10^{-1} cm^{-2} s^{-1}. Total fluence rate for all 2490 lines calculated is 192 cm^{-2} s^{-1}.

E/MeV	$\dot{\Phi}$ /cm^{-2} s^{-1}	Nuclide	E/MeV	$\dot{\Phi}$ /cm^{-2} s^{-1}	Nuclide
5.954E-02	1.135E+02	AM241	1.253E-01	3.107E-01	AM241
4.137E-01	9.385E+00	PU239	3.415E-01	2.923E-01	PU239
3.751E-01	8.278E+00	PU239	9.465E-02	2.686E-01	PU241A
9.843E-02	4.768E+00	PU239	1.113E-01	2.633E-01	PU241A
1.030E-01	3.962E+00	AM241	1.031E-01	2.448E-01	PU239
1.037E-01	3.716E+00	Pu Kα1	5.162E-02	2.294E-01	PU239
9.897E-02	3.147E+00	AM241	3.238E-01	2.131E-01	PU239
1.293E-01	2.885E+00	PU239	3.209E-01	2.102E-01	PU239
3.450E-01	2.507E+00	PU239	2.080E-01	2.081E-01	AM241
1.172E-01	2.410E+00	Pu Kβ1	3.687E-01	2.012E-01	AM241
9.953E-02	2.091E+00	Pu Kα2	6.459E-01	1.971E-01	PU239
3.328E-01	2.069E+00	PU239	2.554E-01	1.881E-01	PU239
3.931E-01	2.040E+00	PU239	1.140E-01	1.854E-01	PU241A
9.465E-02	2.008E+00	PU239	1.486E-01	1.850E-01	PU241A
1.042E-01	2.004E+00	PU240	1.037E-01	1.825E-01	PU241A
1.113E-01	1.981E+00	PU239	1.442E-01	1.694E-01	PU239
3.802E-01	1.668E+00	PU239	2.975E-01	1.642E-01	PU239
3.828E-01	1.435E+00	PU239	9.985E-02	1.633E-01	PU238
4.515E-01	1.413E+00	PU239	4.267E-01	1.576E-01	PU239
9.878E-02	1.245E+00	PU239	3.119E-01	1.523E-01	PA233
3.925E-01	1.191E+00	PU239	9.707E-02	1.485E-01	AM241
1.104E-01	1.028E+00	PU239	1.104E-01	1.378E-01	PU241A
6.624E-01	8.791E-01	AM241	3.450E-01	1.353E-01	PU239
1.163E-01	8.489E-01	PU239	1.957E-01	1.351E-01	PU239
4.226E-01	8.113E-01	PU239	3.767E-01	1.334E-01	AM241
2.036E-01	7.895E-01	PU239	6.190E-01	1.308E-01	AM241
1.144E-01	7.879E-01	PU239	6.589E-01	1.292E-01	PU239
1.154E-01	6.868E-01	PU239	1.142E-01	1.150E-01	AM241
9.843E-02	6.331E-01	PU241A	3.324E-01	1.120E-01	AM241
7.220E-01	5.282E-01	AM241	6.400E-01	1.114E-01	PU239
3.361E-01	4.786E-01	PU239	7.694E-01	1.097E-01	PU239
3.671E-01	4.545E-01	PU239	3.225E-01	1.073E-01	AM241
3.686E-01	4.530E-01	PU239	1.144E-01	1.053E-01	PU241A
3.354E-01	3.799E-01	AM241	1.714E-01	1.007E-01	PU239
1.011E-01	3.283E-01	AM241			

Table 4. Fluence rate from uscattered photons of energy *E* at 1 m in air for the U and Pu spheres.

d. ^{240}Pu (6.0 %) 0.5 kg sphere of age 10 y; fluence rates greater than 10^{-1} cm^{-2} s^{-1}. Total fluence rate for all 2490 lines calculated is 159 cm^{-2} s^{-1}.

E/MeV	$\dot{\Phi}$ /cm^{-2} s^{-1}	Nuclide	*E*/MeV	$\dot{\Phi}$ /cm^{-2} s^{-1}	Nuclide
5.954E-02	8.151E+01	AM241	3.671E-01	4.547E-01	PU239
4.137E-01	9.389E+00	PU239	3.686E-01	4.532E-01	PU239
3.751E-01	8.281E+00	PU239	1.140E-01	3.837E-01	PU241A
9.843E-02	4.770E+00	PU239	1.486E-01	3.829E-01	PU241A
1.037E-01	3.684E+00	Pu Kα1	7.220E-01	3.793E-01	AM241
1.293E-01	2.886E+00	PU239	1.037E-01	3.778E-01	PU241A
1.030E-01	2.845E+00	AM241	3.415E-01	2.925E-01	PU239
3.450E-01	2.508E+00	PU239	1.104E-01	2.852E-01	PU241A
1.172E-01	2.389E+00	Pu Kβ1	3.354E-01	2.728E-01	AM241
9.897E-02	2.260E+00	AM241	1.031E-01	2.449E-01	PU239
9.953E-02	2.073E+00	Pu Kα2	1.011E-01	2.357E-01	AM241
3.328E-01	2.070E+00	PU239	5.162E-02	2.295E-01	PU239
3.931E-01	2.041E+00	PU239	1.253E-01	2.231E-01	AM241
9.465E-02	2.009E+00	PU239	1.144E-01	2.180E-01	PU241A
1.042E-01	2.008E+00	PU240	3.238E-01	2.132E-01	PU239
1.113E-01	1.982E+00	PU239	3.209E-01	2.103E-01	PU239
3.802E-01	1.669E+00	PU239	6.459E-01	1.972E-01	PU239
3.828E-01	1.435E+00	PU239	2.554E-01	1.882E-01	PU239
4.515E-01	1.413E+00	PU239	9.985E-02	1.838E-01	PU238
9.843E-02	1.311E+00	PU241A	1.442E-01	1.695E-01	PU239
9.878E-02	1.246E+00	PU239	2.975E-01	1.643E-01	PU239
3.925E-01	1.192E+00	PU239	4.267E-01	1.577E-01	PU239
1.104E-01	1.029E+00	PU239	2.080E-01	1.494E-01	AM241
1.163E-01	8.493E-01	PU239	3.687E-01	1.445E-01	AM241
4.226E-01	8.116E-01	PU239	3.450E-01	1.353E-01	PU239
2.036E-01	7.898E-01	PU239	1.957E-01	1.352E-01	PU239
1.144E-01	7.882E-01	PU239	6.589E-01	1.292E-01	PU239
1.154E-01	6.871E-01	PU239	6.400E-01	1.115E-01	PU239
6.624E-01	6.313E-01	AM241	7.694E-01	1.097E-01	PU239
9.465E-02	5.561E-01	PU241A	9.707E-02	1.066E-01	AM241
1.113E-01	5.451E-01	PU241A	1.714E-01	1.008E-01	PU239
3.361E-01	4.788E-01	PU239			

Acknowledgements

The author is grateful to Michael Unterweger and Ronald Collé for frequent guidance on the radioactive decay calculations, and to Craig Heimbach who pointed out significant errors in the previous version of this report. This work was supported in part by an Interagency Agreement with the Domestic Nuclear Detection Office of the Department of Homeland Security, IAA HSHQDC-08-X-00277.

References

Bateman, H. (1910). "The solution of a system of differential equations occurring in the theory of radioactive transformations," *Proc. Cambridge Phil. Soc.*, **15**, 423-427.

Berger, M.J., and Hubbell, J.H. (1987). *XCOM: Photon Cross Sections on a Personal Computer*, Report NBSIR 87-3597, National Bureau of Standards, Gaithersburg, MD.

Coursey, J.S., Schwab, D.J., and Dragoset, R.A. (2005). *Atomic Weights and Isotopic Compositions* (version 2.4.1). [Online] Available: http://physics.nist.gov/Comp [2008, April]. National Institute of Standards and Technology, Gaithersburg, MD.

Fetter S., Frolov, V.A., Prilutsky, O.F., and Sagdeev, R.Z. (1990). "Fissile Materials and Weapon Design," Appendix A in *Science & Global Security*, Vol. 1, pp. 225-302, Gordon and Breach Science Publishers.

Halbleib, J.A., Kensek, R.P., Mehlhorn, T.A., Valdez, G.D., Seltzer, S.M., and Berger, M.J. (1992). *ITS Version 3.0: The Integrated TIGER Series of Coupled Electron/Photon Monte Carlo Transport Codes*, Report SAND91-1634, Sandia National Laboratories, Albuquerque, NM.

Hollas, C.L., Goulding, C., and Myers, W. (2005). "Determination of neutron multiplication of subcritical HEU systems using delayed neutrons," Nucl. Inst. and Meth. A, **543**, 559-569.

Larsson, C.L., and Haslip, D.S. (2004). *Consolidated Canadian Results to the HEU Roaund Robin Exercise*, Technical Memorandum 2004-192, Defence Research and Development Canada, Ottawa, Canada.

Nguyen, C.T., and Zsigrai, J. (2006). « Basic characterization of highly enriched unranium by gamma spectrometry," Nucl. Inst. and Meth. B, **246**, 417-424.

Phillips, G.W., Nagel, D.J., and Coffey, T. (2005). *A Primer on the Detection of Nuclear and Radiological Weapons*, Defense & Technology Paper, National Defense University Center for Technology and National Security Policy, Fort Lesley J. McNair, Washington, DC.

Sapr, J.L., Kidman, R, and Brewer, R.W., $^{235}U(94\%)$ *Spheres Surrounded bu Natural-Uranium Reflectors*, Report LA-UR-98-1664 of the Los Alamos National Laboratory, Los

Alamos, NM.

Stefáka, Z, Katoma, R., and Varga, Z. (2008). "Laser ablationb assisted ICP-MS as a tool for rapid categorization od seized unranium oxide materials based on isotopic composition determination," J. Anal. At. Spectrom., **23**, 1030-1033.

Valković, V. (2006). "Applications of nuclear techniques relevant for civil security," J. Phys.: Conf. Series, **41**, 81-100.

Appendix A

The following lists the photon emission data assumed for the relevant radionuclide decay. Photon energies below 1 keV are not shown nor were considered. Radionuclides not listed have no photon emission. Each listing gives the number of lines, followed by a table: of photon energy in MeV, emission probability per decay, and the radionuclide designator (an A or B indicates α or β emission, respectively).

```
   9
1.360000E-02  8.500000E-02  PU242
4.491500E-02  3.730000E-04  PU242
9.465400E-02  1.810000E-07  PU242
9.843400E-02  2.890000E-07  PU242
1.035000E-01  2.550000E-05  PU242
1.104210E-01  3.600000E-08  PU242
1.112980E-01  6.900000E-08  PU242
1.144450E-01  2.660000E-08  PU242
1.588000E-01  3.000000E-06  PU242


  110
1.390000E-02  3.700000E-01  AM241
2.634460E-02  2.270000E-02  AM241
3.319600E-02  1.260000E-03  AM241
4.270400E-02  5.500000E-05  AM241
4.342000E-02  7.300000E-04  AM241
5.101000E-02  2.600000E-07  AM241
5.556000E-02  1.810000E-04  AM241
5.954091E-02  3.590000E-01  AM241
6.483001E-02  1.450000E-06  AM241
6.745000E-02  4.200000E-06  AM241
6.976000E-02  2.900000E-05  AM241
7.580001E-02  5.900000E-06  AM241
9.706900E-02  1.140000E-05  AM241
9.897000E-02  2.030000E-04  AM241
1.010590E-01  1.810000E-05  AM241
1.029800E-01  1.950000E-04  AM241
1.097000E-01  4.900000E-08  AM241
1.133030E-01  2.270000E-06  AM241
1.142340E-01  4.300000E-06  AM241
1.174630E-01  1.680000E-06  AM241
1.203600E-01  4.500000E-08  AM241
1.230520E-01  1.000000E-05  AM241
1.253000E-01  4.080000E-05  AM241
1.394400E-01  5.300000E-08  AM241
1.465500E-01  4.610000E-06  AM241
1.500400E-01  7.400000E-07  AM241
1.542700E-01  5.400000E-09  AM241
1.592600E-01  1.400000E-08  AM241
1.615400E-01  1.500000E-08  AM241
1.646900E-01  6.670000E-07  AM241
1.658100E-01  2.320000E-07  AM241
1.695600E-01  1.730000E-06  AM241
1.750700E-01  1.820000E-07  AM241
1.919600E-01  2.160000E-07  AM241
1.970000E-01  4.900000E-09  AM241
2.040600E-01  2.900000E-08  AM241
2.080100E-01  7.910000E-06  AM241
2.214600E-01  4.240000E-07  AM241
2.328100E-01  4.600000E-08  AM241
2.343300E-01  7.000000E-09  AM241
```

```
2.467300E-01  2.400000E-08  AM241
2.490000E-01  5.400000E-09  AM241
2.608000E-01  1.210000E-08  AM241
2.648900E-01  9.000000E-08  AM241
2.675800E-01  2.630000E-07  AM241
2.757700E-01  6.600000E-08  AM241
2.780400E-01  4.400000E-09  AM241
2.913000E-01  3.100000E-08  AM241
2.927700E-01  1.420000E-07  AM241
3.042100E-01  1.010000E-08  AM241
3.091000E-01  1.400000E-08  AM241
3.168000E-01  3.000000E-10  AM241
3.225200E-01  1.520000E-06  AM241
3.323500E-01  1.490000E-06  AM241
3.353700E-01  4.960000E-06  AM241
3.377000E-01  4.290000E-08  AM241
3.582500E-01  1.200000E-08  AM241
3.686500E-01  2.170000E-06  AM241
3.709400E-01  5.230000E-07  AM241
3.766500E-01  1.380000E-06  AM241
3.838100E-01  2.820000E-07  AM241
3.986400E-01  2.000000E-08  AM241
4.013000E-01  4.900000E-09  AM241
4.063500E-01  1.450000E-08  AM241
4.193300E-01  2.870000E-07  AM241
4.264700E-01  2.460000E-07  AM241
4.464300E-01  4.900000E-09  AM241
4.526000E-01  2.400000E-08  AM241
4.546600E-01  9.700000E-08  AM241
4.596800E-01  3.600000E-08  AM241
4.632200E-01  1.000000E-08  AM241
4.681200E-01  2.880000E-08  AM241
4.873000E-01  4.400000E-09  AM241
5.125000E-01  1.150000E-08  AM241
5.140000E-01  2.600000E-08  AM241
5.220600E-01  1.000000E-08  AM241
5.291700E-01  4.600000E-09  AM241
5.454000E-01  7.400000E-09  AM241
5.630500E-01  7.400000E-09  AM241
5.739400E-01  1.250000E-08  AM241
5.865901E-01  1.310000E-08  AM241
5.902801E-01  2.860000E-08  AM241
5.974800E-01  7.400000E-08  AM241
6.190100E-01  5.940000E-07  AM241
6.271800E-01  5.600000E-09  AM241
6.329300E-01  1.260000E-08  AM241
6.414700E-01  7.100000E-08  AM241
6.530200E-01  3.770000E-07  AM241
6.624001E-01  3.640000E-06  AM241
6.665000E-01  4.900000E-09  AM241
6.698300E-01  3.800000E-09  AM241
6.760300E-01  6.400000E-09  AM241
6.801000E-01  3.130000E-08  AM241
```

6.887200E-01	3.250000E-07	AM241
6.966000E-01	5.340000E-08	AM241
7.220100E-01	1.960000E-06	AM241
7.297200E-01	1.330000E-08	AM241
7.373400E-01	8.000000E-08	AM241
7.559001E-01	7.600000E-08	AM241
7.670000E-01	5.000000E-08	AM241
7.705700E-01	4.740000E-08	AM241
7.724001E-01	2.660000E-08	AM241
7.860000E-01	6.200000E-09	AM241
8.019400E-01	1.360000E-08	AM241
8.062600E-01	3.100000E-09	AM241
8.285000E-01	2.400000E-09	AM241
8.607001E-01	8.000000E-10	AM241
8.627000E-01	5.300000E-09	AM241
8.873000E-01	2.200000E-09	AM241
9.215000E-01	1.900000E-09	AM241

19

1.360000E-02	1.250000E-05	PU241A
2.670000E-02	1.000000E-06	PU241A
4.420000E-02	4.165000E-08	PU241A
4.486000E-02	8.330000E-09	PU241A
5.632000E-02	2.499000E-08	PU241A
5.676000E-02	9.800000E-09	PU241A
7.160001E-02	2.870000E-08	PU241A
7.710000E-02	2.060000E-07	PU241A
9.465400E-02	3.120000E-06	PU241A
9.843400E-02	4.990000E-06	PU241A
1.010000E-01	7.100000E-10	PU241A
1.036800E-01	1.012000E-06	PU241A
1.104210E-01	6.250000E-07	PU241A
1.112980E-01	1.180000E-06	PU241A
1.140000E-01	8.100000E-07	PU241A
1.144450E-01	4.590000E-07	PU241A
1.212000E-01	6.860000E-09	PU241A
1.485670E-01	1.860000E-06	PU241A
1.599550E-01	6.539999E-08	PU241A

18

1.360000E-02	9.600000E-02	PU240
4.524400E-02	4.470000E-04	PU240
9.465400E-02	2.540000E-07	PU240
9.843400E-02	4.050000E-07	PU240
1.042340E-01	7.140000E-05	PU240
1.104210E-01	5.080000E-08	PU240
1.112980E-01	9.600000E-08	PU240
1.144450E-01	3.730000E-08	PU240
1.603080E-01	4.020000E-06	PU240
2.124600E-01	2.900000E-07	PU240
5.380901E-01	1.470000E-09	PU240
6.423500E-01	1.300000E-07	PU240
6.875700E-01	3.500000E-08	PU240
6.990001E-01	1.300000E-10	PU240
8.739200E-01	5.800000E-09	PU240
9.580001E-01	5.000000E-10	PU240
9.600000E-01	3.000000E-10	PU240
9.670001E-01	3.000000E-10	PU240

174

1.297500E-02	3.410000E-04	PU239
1.360000E-02	4.380000E-02	PU239
3.004000E-02	2.170000E-06	PU239
3.866100E-02	1.044000E-04	PU239
4.193000E-02	1.460000E-06	PU239
4.621000E-02	7.210000E-07	PU239
4.668000E-02	4.650000E-07	PU239
4.760000E-02	6.250000E-07	PU239
5.162400E-02	2.722000E-04	PU239
5.403901E-02	1.944000E-06	PU239
5.682800E-02	1.152000E-05	PU239
6.570800E-02	5.200000E-07	PU239
6.767400E-02	1.517000E-06	PU239
6.869600E-02	3.600000E-06	PU239
6.874000E-02	1.300000E-06	PU239
7.759201E-02	3.800000E-06	PU239
7.843000E-02	1.542000E-06	PU239
8.964001E-02	2.700000E-07	PU239
9.465400E-02	3.600000E-05	PU239
9.614000E-02	3.790000E-07	PU239
9.760001E-02	8.000000E-07	PU239
9.843400E-02	5.800000E-05	PU239
9.878001E-02	1.470000E-05	PU239
1.030600E-01	2.160000E-06	PU239
1.104210E-01	7.200000E-06	PU239
1.112980E-01	1.370000E-05	PU239
1.144450E-01	5.300000E-06	PU239
1.153800E-01	4.600000E-06	PU239
1.162600E-01	5.670000E-06	PU239
1.197000E-01	2.100000E-07	PU239
1.223500E-01	9.500000E-09	PU239
1.232280E-01	1.600000E-11	PU239
1.236200E-01	2.370000E-07	PU239
1.245100E-01	6.810000E-07	PU239
1.252100E-01	5.630000E-07	PU239
1.292960E-01	6.310000E-05	PU239
1.416570E-01	3.200000E-07	PU239
1.433500E-01	1.730000E-07	PU239
1.442010E-01	2.830000E-06	PU239
1.460940E-01	1.190000E-06	PU239
1.581000E-01	1.000000E-08	PU239
1.601900E-01	6.200000E-08	PU239
1.614500E-01	1.230000E-06	PU239
1.678100E-01	2.900000E-08	PU239
1.713930E-01	1.100000E-06	PU239
1.725600E-01	3.000000E-11	PU239
1.737000E-01	3.100000E-08	PU239
1.792200E-01	6.600000E-07	PU239
1.882300E-01	1.090000E-07	PU239
1.893600E-01	8.300000E-07	PU239
1.956790E-01	1.070000E-06	PU239
2.035500E-01	5.690000E-06	PU239
2.180000E-01	1.200000E-08	PU239
2.254200E-01	1.510000E-07	PU239
2.377700E-01	1.440000E-07	PU239
2.420800E-01	7.300000E-08	PU239
2.433800E-01	2.530000E-07	PU239
2.449200E-01	5.100000E-08	PU239
2.489500E-01	7.200000E-08	PU239
2.553840E-01	8.000000E-07	PU239
2.639500E-01	2.650000E-07	PU239
2.657000E-01	1.600000E-08	PU239
2.812000E-01	2.100000E-08	PU239
2.853000E-01	1.900000E-08	PU239
2.974600E-01	4.980000E-07	PU239
3.028700E-01	5.100000E-08	PU239

3.078500E-01 5.500000E-08 PU239
3.117800E-01 2.580000E-07 PU239
3.164100E-01 1.320000E-07 PU239
3.196800E-01 4.800000E-08 PU239
3.208620E-01 5.420000E-07 PU239
3.238400E-01 5.390000E-07 PU239
3.328450E-01 4.940000E-06 PU239
3.361130E-01 1.120000E-06 PU239
3.415060E-01 6.620000E-07 PU239
3.450130E-01 5.560000E-06 PU239
3.450140E-01 3.000000E-07 PU239
3.540000E-01 7.000000E-09 PU239
3.618900E-01 1.220000E-07 PU239
3.670730E-01 8.900000E-07 PU239
3.685540E-01 8.800000E-07 PU239
3.750540E-01 1.554000E-05 PU239
3.801910E-01 3.050000E-06 PU239
3.827500E-01 2.590000E-06 PU239
3.925300E-01 2.050000E-06 PU239
3.931400E-01 3.500000E-06 PU239
3.995300E-01 5.900000E-08 PU239
4.068000E-01 2.500000E-08 PU239
4.112000E-01 7.000000E-08 PU239
4.123000E-01 1.800000E-10 PU239
4.137130E-01 1.466000E-05 PU239
4.225980E-01 1.220000E-06 PU239
4.266800E-01 2.330000E-07 PU239
4.284000E-01 1.000000E-08 PU239
4.300800E-01 4.300000E-08 PU239
4.457200E-01 8.799999E-08 PU239
4.514810E-01 1.894000E-06 PU239
4.576100E-01 1.490000E-08 PU239
4.612500E-01 2.270000E-08 PU239
4.639000E-01 2.800000E-09 PU239
4.739000E-01 5.000000E-10 PU239
4.816600E-01 4.600000E-08 PU239
4.870600E-01 2.650000E-09 PU239
4.930800E-01 8.700000E-09 PU239
5.264000E-01 5.700000E-10 PU239
5.505000E-01 4.200000E-09 PU239
5.794001E-01 8.600000E-10 PU239
5.828900E-01 6.150000E-09 PU239
5.863000E-01 1.530000E-09 PU239
5.960000E-01 3.900000E-10 PU239
5.979900E-01 1.670000E-08 PU239
5.996000E-01 2.000000E-09 PU239
6.069000E-01 1.200000E-09 PU239
6.128300E-01 9.500000E-09 PU239
6.171000E-01 1.340000E-08 PU239
6.182801E-01 2.040000E-08 PU239
6.192101E-01 1.210000E-08 PU239
6.247801E-01 4.370000E-09 PU239
6.331500E-01 2.530000E-08 PU239
6.377000E-01 2.560000E-08 PU239
6.378000E-01 2.560000E-08 PU239
6.399900E-01 8.700000E-08 PU239
6.459400E-01 1.520000E-07 PU239
6.493201E-01 7.100000E-09 PU239
6.520500E-01 6.600000E-08 PU239
6.548800E-01 2.250000E-08 PU239
6.588600E-01 9.700000E-08 PU239
6.645800E-01 1.660000E-08 PU239
6.682000E-01 3.900000E-10 PU239

6.708000E-01 8.999999E-11 PU239
6.709900E-01 8.999999E-11 PU239
6.740500E-01 5.150000E-09 PU239
6.744000E-01 5.150000E-09 PU239
6.908100E-01 9.000000E-09 PU239
6.932001E-01 2.000000E-10 PU239
6.978000E-01 7.400000E-10 PU239
7.011000E-01 5.120000E-09 PU239
7.036800E-01 3.950000E-08 PU239
7.147101E-01 7.900000E-10 PU239
7.180001E-01 2.800000E-08 PU239
7.203000E-01 4.900000E-10 PU239
7.279001E-01 1.240000E-09 PU239
7.365000E-01 3.000000E-10 PU239
7.474000E-01 8.100000E-10 PU239
7.564000E-01 3.470000E-08 PU239
7.626000E-01 9.999999E-11 PU239
7.636000E-01 2.200000E-10 PU239
7.664700E-01 1.300000E-09 PU239
7.672900E-01 1.400000E-09 PU239
7.691501E-01 5.100000E-08 PU239
7.693700E-01 6.800000E-08 PU239
7.695400E-01 8.000000E-10 PU239
7.794001E-01 1.360000E-09 PU239
7.929001E-01 2.000000E-10 PU239
8.059000E-01 2.700000E-10 PU239
8.084000E-01 1.210000E-09 PU239
8.137001E-01 4.500000E-10 PU239
8.160000E-01 2.400000E-10 PU239
8.213000E-01 5.500000E-10 PU239
8.325000E-01 2.960000E-10 PU239
8.404000E-01 4.800000E-10 PU239
8.437800E-01 1.340000E-09 PU239
8.792000E-01 3.600000E-10 PU239
8.910000E-01 7.500000E-10 PU239
9.187000E-01 7.999999E-11 PU239
9.403000E-01 5.000000E-10 PU239
9.556000E-01 3.100000E-10 PU239
9.576000E-01 3.200000E-10 PU239
9.683700E-01 2.800000E-10 PU239
9.797001E-01 2.800000E-10 PU239
9.869001E-01 2.100000E-10 PU239
9.927000E-01 2.700000E-10 PU239
1.005700E+00 1.800000E-10 PU239
1.057300E+00 4.500000E-10 PU239

36
1.360000E-02 1.020000E-01 PU238
4.349800E-02 3.920000E-04 PU238
6.270000E-02 5.000000E-11 PU238
9.465400E-02 5.790000E-07 PU238
9.843400E-02 9.249999E-07 PU238
9.985300E-02 7.290000E-05 PU238
1.104210E-01 1.160000E-07 PU238
1.112980E-01 2.200000E-07 PU238
1.144450E-01 8.520000E-08 PU238
1.401500E-01 9.999999E-12 PU238
1.527200E-01 9.289999E-06 PU238
1.929100E-01 6.400000E-12 PU238
2.009700E-01 3.900000E-08 PU238
2.031200E-01 4.000000E-11 PU238
2.359000E-01 9.999999E-11 PU238
2.583000E-01 8.400000E-10 PU238

2.992000E-01 4.800000E-10 PU238
7.059001E-01 5.300000E-10 PU238
7.084200E-01 4.100000E-09 PU238
7.428100E-01 5.200000E-08 PU238
7.663900E-01 2.200000E-07 PU238
7.834001E-01 2.400000E-10 PU238
7.863000E-01 3.200000E-08 PU238
8.044001E-01 1.200000E-09 PU238
8.056000E-01 5.999999E-10 PU238
8.082500E-01 7.900000E-09 PU238
8.517001E-01 1.240000E-08 PU238
8.805000E-01 1.600000E-09 PU238
8.832300E-01 7.600000E-09 PU238
9.043000E-01 6.399999E-10 PU238
9.267200E-01 5.800000E-09 PU238
9.419001E-01 4.700000E-09 PU238
9.460000E-01 4.000000E-10 PU238
1.001030E+00 9.800000E-09 PU238
1.041800E+00 2.200000E-09 PU238
1.085400E+00 9.100000E-10 PU238

    8
1.300000E-02 7.300000E-02 U238
4.955000E-02 6.400000E-04 U238
8.995701E-02 6.800000E-06 U238
9.335000E-02 1.100000E-05 U238
1.048190E-01 1.360000E-06 U238
1.056040E-01 2.600000E-06 U238
1.085820E-01 9.900000E-07 U238
1.135000E-01 1.020000E-04 U238

    28
1.381000E-02 9.900000E-04 U237
1.390000E-02 6.200000E-01 U237
2.634460E-02 2.430000E-02 U237
3.319600E-02 1.300000E-03 U237
3.854000E-02 4.000000E-03 U237
4.342000E-02 2.400000E-04 U237
5.101000E-02 3.400000E-03 U237
5.954091E-02 3.450000E-01 U237
6.483001E-02 1.282000E-02 U237
6.976000E-02 9.500000E-06 U237
9.706900E-02 1.540000E-01 U237
1.010590E-01 2.450000E-01 U237
1.029800E-01 6.400000E-05 U237
1.133030E-01 3.070000E-02 U237
1.142340E-01 5.810000E-02 U237
1.174630E-01 2.270000E-02 U237
1.646100E-01 1.860000E-02 U237
2.080050E-01 2.120000E-01 U237
2.218000E-01 2.120000E-04 U237
2.344000E-01 2.050000E-04 U237
2.675400E-01 7.120000E-03 U237
2.927700E-01 2.500000E-05 U237
3.091000E-01 2.700000E-06 U237
3.323500E-01 1.200000E-02 U237
3.353700E-01 9.510000E-04 U237
3.377000E-01 8.899999E-05 U237
3.686200E-01 3.920000E-04 U237
3.709400E-01 1.073000E-03 U237

    54
5.180000E-03 2.220000E-03 NP237
8.220000E-03 9.000000E-02 NP237
1.330000E-02 4.930000E-01 NP237
2.937400E-02 1.412000E-01 NP237
3.632000E-02 5.000000E-05 NP237
4.653000E-02 1.040000E-03 NP237
5.710400E-02 3.540000E-03 NP237
6.259000E-02 6.000000E-05 NP237
6.390000E-02 1.080000E-04 NP237
7.049000E-02 1.080000E-04 NP237
7.454000E-02 1.200000E-04 NP237
8.647700E-02 1.240000E-01 NP237
8.799000E-02 1.670000E-03 NP237
9.228200E-02 1.660000E-02 NP237
9.464000E-02 6.150000E-03 NP237
9.586301E-02 2.680000E-02 NP237
1.061500E-01 4.899999E-04 NP237
1.075950E-01 3.330000E-03 NP237
1.084220E-01 6.300000E-03 NP237
1.087000E-01 6.800000E-04 NP237
1.114860E-01 2.440000E-03 NP237
1.154000E-01 2.600000E-05 NP237
1.177020E-01 1.690000E-03 NP237
1.311010E-01 8.569999E-04 NP237
1.342850E-01 6.700000E-04 NP237
1.399000E-01 4.600000E-05 NP237
1.432490E-01 4.430000E-03 NP237
1.514140E-01 2.320000E-03 NP237
1.533700E-01 6.100000E-05 NP237
1.552390E-01 8.890000E-04 NP237
1.624100E-01 3.270000E-04 NP237
1.691560E-01 6.330000E-04 NP237
1.705900E-01 2.000000E-04 NP237
1.761200E-01 1.200000E-04 NP237
1.808100E-01 1.580000E-04 NP237
1.868600E-01 3.000000E-05 NP237
1.914600E-01 1.920000E-04 NP237
1.932600E-01 4.370000E-04 NP237
1.946700E-01 3.300000E-04 NP237
1.949500E-01 1.770000E-03 NP237
1.968600E-01 2.080000E-04 NP237
1.999500E-01 5.300000E-05 NP237
2.016200E-01 3.930000E-04 NP237
2.029000E-01 4.800000E-05 NP237
2.091900E-01 1.420000E-04 NP237
2.122900E-01 1.510000E-03 NP237
2.140100E-01 3.620000E-04 NP237
2.226000E-01 2.000000E-05 NP237
2.299400E-01 1.100000E-04 NP237
2.378600E-01 5.690000E-04 NP237
2.489500E-01 5.000000E-05 NP237
2.570900E-01 6.400000E-05 NP237
2.624400E-01 4.710000E-05 NP237
2.796500E-01 1.090000E-04 NP237

    9
1.300000E-02 9.000000E-02 U236
4.946000E-02 7.800000E-04 U236
8.995701E-02 1.240000E-05 U236
9.335000E-02 2.000000E-05 U236
1.048190E-01 2.500000E-06 U236
1.056040E-01 4.700000E-06 U236

26

1.085820E-01	1.800000E-06	U236
1.127900E-01	1.900000E-04	U236
1.711500E-01	6.200000E-07	U236

57

1.300000E-02	2.800000E-01	U235
1.959000E-02	6.000000E-05	U235
3.160000E-02	1.700000E-04	U235
3.470000E-02	3.700000E-04	U235
4.140000E-02	3.000000E-04	U235
4.196000E-02	6.000000E-04	U235
5.122000E-02	3.400000E-04	U235
5.410000E-02	8.999999E-06	U235
5.425000E-02	1.500000E-04	U235
6.435000E-02	1.300000E-04	U235
7.270000E-02	1.100000E-03	U235
7.502000E-02	6.000000E-04	U235
7.619800E-02	8.000001E-05	U235
8.995701E-02	3.470000E-02	U235
9.335000E-02	5.600000E-02	U235
9.609000E-02	9.099999E-04	U235
1.048190E-01	6.930000E-03	U235
1.056040E-01	1.320000E-02	U235
1.085820E-01	5.060000E-03	U235
1.091600E-01	1.540000E-02	U235
1.154500E-01	3.000000E-04	U235
1.203500E-01	2.600000E-04	U235
1.365500E-01	1.200000E-04	U235
1.407600E-01	2.200000E-03	U235
1.424000E-01	5.000000E-05	U235
1.437600E-01	1.096000E-01	U235
1.509300E-01	8.000000E-04	U235
1.633300E-01	5.080000E-02	U235
1.733000E-01	6.000000E-05	U235
1.826100E-01	3.400000E-03	U235
1.857150E-01	5.720000E-01	U235
1.949400E-01	6.300000E-03	U235
1.989000E-01	4.200000E-04	U235
2.021100E-01	1.080000E-02	U235
2.053110E-01	5.010000E-02	U235
2.152800E-01	2.900000E-04	U235
2.213800E-01	1.200000E-03	U235
2.287800E-01	7.000000E-05	U235
2.335000E-01	2.900000E-04	U235
2.408700E-01	7.500000E-04	U235
2.468400E-01	5.300000E-04	U235
2.664500E-01	6.000000E-05	U235
2.751290E-01	5.200000E-04	U235
2.754300E-01	7.000000E-05	U235
2.814200E-01	6.000000E-05	U235
2.829200E-01	6.000000E-05	U235
2.895600E-01	7.000000E-05	U235
2.916500E-01	4.000000E-04	U235
3.017000E-01	5.000000E-05	U235
3.171000E-01	1.000000E-05	U235
3.435000E-01	3.000000E-05	U235
3.459000E-01	4.000000E-04	U235
3.560300E-01	5.000000E-05	U235
3.878200E-01	4.000000E-04	U235
4.102900E-01	3.000000E-05	U235
4.287100E-01	1.000000E-05	U235
4.484000E-01	1.000000E-05	U235

15

1.300000E-02	9.999999E-02	U234
5.320000E-02	1.230000E-03	U234
8.995701E-02	2.510000E-05	U234
9.335000E-02	4.050000E-05	U234
1.048190E-01	5.010000E-06	U234
1.056040E-01	9.550000E-06	U234
1.085820E-01	3.660000E-06	U234
1.209000E-01	4.000000E-04	U234
4.549500E-01	2.500000E-07	U234
5.035000E-01	9.599999E-09	U234
5.082000E-01	1.500000E-07	U234
5.817000E-01	1.200000E-07	U234
6.244001E-01	8.000000E-09	U234
6.349000E-01	1.400000E-07	U234
6.776000E-01	9.800000E-09	U234

124

1.360000E-02	1.340000E-04	PA234M
6.270000E-02	1.200000E-05	PA234M
9.465400E-02	6.600000E-05	PA234M
9.843400E-02	1.060000E-04	PA234M
1.104210E-01	1.330000E-05	PA234M
1.112980E-01	2.500000E-05	PA234M
1.144450E-01	9.699999E-06	PA234M
1.353200E-01	4.300000E-08	PA234M
1.372300E-01	4.700000E-07	PA234M
1.401000E-01	1.280000E-05	PA234M
1.665000E-01	2.400000E-09	PA234M
1.847000E-01	1.680000E-05	PA234M
1.934000E-01	1.440000E-05	PA234M
1.979100E-01	2.700000E-07	PA234M
1.999000E-01	5.700000E-06	PA234M
2.033000E-01	1.700000E-05	PA234M
2.099000E-01	1.350000E-05	PA234M
2.359000E-01	8.000000E-07	PA234M
2.477000E-01	2.440000E-06	PA234M
2.582270E-01	7.639999E-04	PA234M
2.755000E-01	3.100000E-06	PA234M
2.990000E-01	6.500000E-06	PA234M
3.110000E-01	8.400000E-07	PA234M
3.167000E-01	1.900000E-06	PA234M
3.381000E-01	1.100000E-05	PA234M
3.402000E-01	7.000000E-07	PA234M
3.575000E-01	8.000000E-06	PA234M
3.628000E-01	6.800000E-06	PA234M
3.876000E-01	1.398000E-05	PA234M
4.274000E-01	2.000000E-07	PA234M
4.459100E-01	3.000000E-07	PA234M
4.509700E-01	3.110000E-05	PA234M
4.535800E-01	2.130000E-05	PA234M
4.567000E-01	7.200000E-06	PA234M
4.684300E-01	2.290000E-05	PA234M
4.757400E-01	2.390000E-05	PA234M
4.854400E-01	1.900000E-07	PA234M
5.075001E-01	1.570000E-05	PA234M
5.092000E-01	2.100000E-05	PA234M
5.166000E-01	1.300000E-07	PA234M
5.260201E-01	9.300000E-08	PA234M
5.439800E-01	3.670000E-05	PA234M
5.572400E-01	8.400000E-08	PA234M
5.720000E-01	8.670000E-06	PA234M
5.811900E-01	8.200000E-07	PA234M

6.246000E-01 1.150000E-06 PA234M
6.490000E-01 1.041000E-05 PA234M
6.553000E-01 1.380000E-05 PA234M
6.708000E-01 3.700000E-06 PA234M
6.739001E-01 6.500000E-06 PA234M
6.834000E-01 5.700000E-06 PA234M
6.910000E-01 8.920000E-05 PA234M
6.955000E-01 1.620000E-05 PA234M
6.990200E-01 5.700000E-05 PA234M
7.020000E-01 7.240000E-05 PA234M
7.059400E-01 5.600000E-05 PA234M
7.082000E-01 6.730000E-06 PA234M
7.190101E-01 2.600000E-07 PA234M
7.325000E-01 1.300000E-05 PA234M
7.401000E-01 1.090000E-04 PA234M
7.428131E-01 1.066000E-03 PA234M
7.501200E-01 1.800000E-07 PA234M
7.605301E-01 4.300000E-08 PA234M
7.664200E-01 3.170000E-03 PA234M
7.817500E-01 7.770000E-05 PA234M
7.834001E-01 3.900000E-07 PA234M
7.862801E-01 5.440000E-04 PA234M
7.919400E-01 9.999999E-08 PA234M
8.057500E-01 5.900000E-05 PA234M
8.082001E-01 2.780000E-05 PA234M
8.182001E-01 1.000000E-05 PA234M
8.256000E-01 6.600000E-05 PA234M
8.441000E-01 1.100000E-05 PA234M
8.515801E-01 6.899999E-05 PA234M
8.668000E-01 1.140000E-05 PA234M
8.809001E-01 4.000000E-05 PA234M
8.832200E-01 3.500000E-05 PA234M
9.217200E-01 1.278000E-04 PA234M
9.266101E-01 1.240000E-05 PA234M
9.363000E-01 1.100000E-05 PA234M
9.419601E-01 2.520000E-05 PA234M
9.459400E-01 1.010000E-04 PA234M
9.600000E-01 8.000000E-06 PA234M
9.961000E-01 5.600000E-05 PA234M
1.001030E+00 8.420000E-03 PA234M
1.041700E+00 1.240000E-05 PA234M
1.059400E+00 2.280000E-05 PA234M
1.061860E+00 2.140000E-05 PA234M
1.081900E+00 8.999999E-06 PA234M
1.084250E+00 8.000000E-06 PA234M
1.120600E+00 1.720000E-05 PA234M
1.124930E+00 3.370000E-05 PA234M
1.174200E+00 1.940000E-05 PA234M
1.193730E+00 1.358000E-04 PA234M
1.237260E+00 5.250000E-05 PA234M
1.392700E+00 3.460000E-05 PA234M
1.413880E+00 2.260000E-05 PA234M
1.434140E+00 9.730000E-05 PA234M
1.458500E+00 1.900000E-05 PA234M
1.501000E+00 9.300000E-06 PA234M
1.510210E+00 1.305000E-04 PA234M
1.527270E+00 2.010000E-05 PA234M
1.550000E+00 1.140000E-05 PA234M
1.553750E+00 8.209999E-05 PA234M
1.558400E+00 7.200000E-06 PA234M
1.570670E+00 1.090000E-05 PA234M
1.593870E+00 1.890000E-05 PA234M
1.601800E+00 4.700000E-06 PA234M

1.667600E+00 1.150000E-05 PA234M
1.694100E+00 4.500000E-06 PA234M
1.737750E+00 2.130000E-04 PA234M
1.765440E+00 8.750000E-05 PA234M
1.796200E+00 4.200000E-06 PA234M
1.809040E+00 3.740000E-05 PA234M
1.819690E+00 8.800000E-06 PA234M
1.831360E+00 1.742000E-04 PA234M
1.863090E+00 1.210000E-05 PA234M
1.867690E+00 9.259999E-05 PA234M
1.874880E+00 8.180000E-05 PA234M
1.893510E+00 1.890000E-05 PA234M
1.911190E+00 6.240000E-05 PA234M
1.926500E+00 4.500000E-06 PA234M
1.937040E+00 2.100000E-05 PA234M
1.970000E+00 4.100000E-06 PA234M

16
1.330000E-02 6.900000E-02 TH234
2.002000E-02 5.000000E-05 TH234
2.949000E-02 1.200000E-05 TH234
6.286000E-02 1.600000E-04 TH234
6.329001E-02 3.700000E-02 TH234
7.392000E-02 1.300000E-04 TH234
8.330001E-02 6.000000E-04 TH234
9.228200E-02 1.710000E-04 TH234
9.238000E-02 2.130000E-02 TH234
9.280001E-02 2.100000E-02 TH234
9.586301E-02 2.800000E-04 TH234
1.033500E-01 3.200000E-05 TH234
1.075950E-01 3.400000E-05 TH234
1.084220E-01 6.500000E-05 TH234
1.114860E-01 2.500000E-05 TH234
1.128100E-01 2.100000E-03 TH234

107
1.300000E-02 5.500000E-02 U233
2.531800E-02 1.110000E-05 U233
2.919200E-02 1.200000E-04 U233
3.152000E-02 2.500000E-06 U233
3.240000E-02 9.100000E-06 U233
3.798000E-02 3.300000E-06 U233
4.244000E-02 8.620000E-04 U233
5.262000E-02 2.300000E-06 U233
5.360800E-02 4.100000E-05 U233
5.469900E-02 1.820000E-04 U233
6.388000E-02 3.000000E-07 U233
6.612200E-02 7.700000E-06 U233
6.794301E-02 2.900000E-06 U233
6.887001E-02 9.800000E-07 U233
7.028000E-02 5.500000E-06 U233
7.181900E-02 2.400000E-05 U233
7.288000E-02 5.400000E-06 U233
7.457000E-02 1.500000E-05 U233
7.639001E-02 3.600000E-06 U233
7.713000E-02 6.600000E-06 U233
7.815000E-02 5.500000E-07 U233
8.295701E-02 1.600000E-06 U233
8.543000E-02 1.700000E-06 U233
8.677000E-02 1.200000E-06 U233
8.727001E-02 1.700000E-06 U233
8.846001E-02 4.000000E-06 U233
8.995701E-02 8.130000E-05 U233

9.103000E-02 3.000000E-06 U233
9.335000E-02 1.310000E-04 U233
9.624401E-02 1.270000E-05 U233
9.713401E-02 2.000000E-04 U233
1.000300E-01 5.000000E-07 U233
1.017700E-01 8.200000E-07 U233
1.036000E-01 9.200000E-07 U233
1.048190E-01 1.620000E-05 U233
1.056040E-01 3.100000E-05 U233
1.085820E-01 1.190000E-05 U233
1.120000E-01 4.500000E-06 U233
1.144000E-01 2.300000E-06 U233
1.164100E-01 1.900000E-06 U233
1.171590E-01 2.300000E-05 U233
1.189680E-01 4.060000E-05 U233
1.208160E-01 3.320000E-05 U233
1.238930E-01 5.900000E-06 U233
1.254100E-01 6.000000E-07 U233
1.292500E-01 6.400000E-07 U233
1.312000E-01 3.000000E-07 U233
1.353600E-01 2.320000E-05 U233
1.397600E-01 9.600000E-07 U233
1.444000E-01 2.700000E-06 U233
1.453370E-01 1.500000E-05 U233
1.463450E-01 6.570000E-05 U233
1.481560E-01 3.300000E-06 U233
1.531000E-01 5.000000E-07 U233
1.547700E-01 1.400000E-06 U233
1.561400E-01 5.300000E-07 U233
1.624000E-01 6.900000E-07 U233
1.645220E-01 6.230000E-05 U233
1.657000E-01 3.500000E-06 U233
1.690100E-01 6.200000E-07 U233
1.708400E-01 1.300000E-06 U233
1.723600E-01 3.200000E-07 U233
1.741900E-01 2.100000E-06 U233
1.843000E-01 2.300000E-07 U233
1.879690E-01 1.900000E-05 U233
1.921300E-01 3.700000E-07 U233
2.060000E-01 6.000000E-07 U233
2.081710E-01 2.290000E-05 U233
2.123400E-01 1.260000E-06 U233
2.160800E-01 6.100000E-06 U233
2.171590E-01 3.200000E-05 U233
2.177000E-01 4.600000E-07 U233
2.193800E-01 1.400000E-06 U233
2.233000E-01 3.000000E-07 U233
2.267000E-01 9.000000E-08 U233
2.281000E-01 2.000000E-07 U233
2.301100E-01 6.200000E-07 U233
2.403900E-01 3.500000E-06 U233
2.453450E-01 3.620000E-05 U233
2.487260E-01 1.430000E-05 U233
2.559400E-01 3.900000E-07 U233
2.593300E-01 1.600000E-06 U233
2.606500E-01 9.800000E-07 U233
2.619200E-01 2.800000E-06 U233
2.686600E-01 2.300000E-06 U233
2.723400E-01 5.700000E-07 U233
2.747280E-01 4.000000E-06 U233
2.781110E-01 1.080000E-05 U233
2.880330E-01 9.699999E-06 U233
2.913540E-01 5.370000E-05 U233

2.939100E-01 1.300000E-06 U233
3.028900E-01 6.400000E-07 U233
3.095000E-01 6.600000E-07 U233
3.120000E-01 2.500000E-07 U233
3.171600E-01 7.759999E-05 U233
3.205410E-01 2.900000E-05 U233
3.234200E-01 7.700000E-06 U233
3.285370E-01 6.000000E-07 U233
3.366100E-01 5.400000E-06 U233
3.389000E-01 7.000000E-08 U233
3.540300E-01 5.300000E-07 U233
3.657900E-01 7.500000E-06 U233
3.834700E-01 8.700000E-07 U233
3.937000E-01 7.000000E-08 U233
3.967000E-01 8.000000E-08 U233
4.366000E-01 4.600000E-08 U233
4.841000E-01 4.000000E-08 U233

190
1.360000E-02 4.300000E-01 PA233
1.726000E-02 4.000000E-05 PA233
1.870000E-02 4.600000E-04 PA233
1.970000E-02 4.600000E-04 PA233
2.390000E-02 3.100000E-05 PA233
2.470000E-02 3.100000E-05 PA233
2.855900E-02 7.100000E-04 PA233
3.190000E-02 2.300000E-05 PA233
3.530000E-02 1.500000E-05 PA233
3.580000E-02 1.900000E-05 PA233
3.850000E-02 6.300000E-05 PA233
4.034900E-02 2.360000E-04 PA233
4.166300E-02 1.400000E-04 PA233
4.170000E-02 1.000000E-04 PA233
4.270000E-02 1.900000E-05 PA233
4.580000E-02 4.000000E-06 PA233
4.670000E-02 8.000000E-06 PA233
4.770000E-02 8.000000E-06 PA233
4.880000E-02 1.600000E-05 PA233
4.970000E-02 8.000000E-06 PA233
5.180000E-02 3.600000E-05 PA233
5.250000E-02 1.600000E-05 PA233
5.320000E-02 1.200000E-05 PA233
5.500000E-02 8.000000E-06 PA233
5.800000E-02 1.200000E-05 PA233
5.920000E-02 8.000000E-06 PA233
5.960000E-02 8.000000E-06 PA233
6.060000E-02 2.400000E-04 PA233
6.160000E-02 1.600000E-05 PA233
6.320000E-02 8.000000E-06 PA233
6.360000E-02 8.000000E-06 PA233
6.550001E-02 2.700000E-05 PA233
6.640001E-02 1.900000E-05 PA233
6.750000E-02 1.900000E-05 PA233
6.850000E-02 2.700000E-05 PA233
6.960000E-02 4.600000E-05 PA233
7.030001E-02 2.700000E-05 PA233
7.130001E-02 7.800000E-05 PA233
7.400000E-02 3.500000E-05 PA233
7.526900E-02 1.320000E-02 PA233
7.700000E-02 3.800000E-05 PA233
7.790001E-02 1.200000E-05 PA233
7.840001E-02 7.700000E-05 PA233
7.910001E-02 8.000000E-06 PA233

29

8.080000E-02	1.500000E-05	PA233
8.180001E-02	4.500000E-05	PA233
8.250000E-02	1.500000E-05	PA233
8.480000E-02	7.000000E-05	PA233
8.520000E-02	1.400000E-04	PA233
8.659501E-02	1.950000E-02	PA233
8.900000E-02	7.000000E-05	PA233
8.930001E-02	7.000000E-05	PA233
9.000000E-02	1.200000E-05	PA233
9.100001E-02	1.200000E-05	PA233
9.150001E-02	1.200000E-05	PA233
9.220000E-02	7.000000E-05	PA233
9.270000E-02	1.200000E-05	PA233
9.300000E-02	8.000000E-06	PA233
9.350000E-02	8.000000E-06	PA233
9.465400E-02	1.056000E-01	PA233
9.670001E-02	4.000000E-05	PA233
9.700000E-02	5.000000E-05	PA233
9.843400E-02	1.690000E-01	PA233
1.006000E-01	3.100000E-05	PA233
1.021000E-01	1.900000E-05	PA233
1.025000E-01	4.600000E-05	PA233
1.038600E-01	8.540000E-03	PA233
1.057000E-01	8.000000E-06	PA233
1.063000E-01	1.600000E-05	PA233
1.081000E-01	1.200000E-05	PA233
1.104210E-01	2.120000E-02	PA233
1.112980E-01	4.010000E-02	PA233
1.130000E-01	3.500000E-05	PA233
1.144450E-01	1.550000E-02	PA233
1.165000E-01	5.800000E-05	PA233
1.169000E-01	5.800000E-05	PA233
1.196000E-01	8.000000E-06	PA233
1.220000E-01	1.500000E-05	PA233
1.251000E-01	3.000000E-05	PA233
1.283000E-01	1.200000E-05	PA233
1.300000E-01	1.200000E-05	PA233
1.310000E-01	3.000000E-05	PA233
1.329000E-01	1.200000E-05	PA233
1.352000E-01	4.600000E-05	PA233
1.358000E-01	1.500000E-05	PA233
1.365000E-01	1.900000E-05	PA233
1.393000E-01	4.600000E-05	PA233
1.427000E-01	2.300000E-05	PA233
1.431000E-01	1.500000E-05	PA233
1.444000E-01	7.000000E-05	PA233
1.485000E-01	2.700000E-05	PA233
1.505000E-01	2.300000E-05	PA233
1.537000E-01	3.900000E-05	PA233
1.547000E-01	4.600000E-05	PA233
1.561000E-01	2.300000E-05	PA233
1.570000E-01	2.700000E-05	PA233
1.579000E-01	2.300000E-05	PA233
1.591000E-01	3.900000E-05	PA233
1.600000E-01	3.100000E-05	PA233
1.612000E-01	2.700000E-05	PA233
1.624000E-01	2.300000E-05	PA233
1.633000E-01	2.300000E-05	PA233
1.666000E-01	1.200000E-05	PA233
1.680000E-01	3.100000E-05	PA233
1.706000E-01	5.000000E-05	PA233
1.728000E-01	2.700000E-05	PA233
1.737000E-01	8.399999E-05	PA233
1.747000E-01	4.200000E-05	PA233
1.752000E-01	1.200000E-05	PA233
1.780000E-01	5.400000E-05	PA233
1.801000E-01	2.700000E-05	PA233
1.827000E-01	2.700000E-05	PA233
1.833000E-01	1.200000E-05	PA233
1.848000E-01	3.100000E-05	PA233
1.857000E-01	3.500000E-05	PA233
1.985000E-01	3.100000E-05	PA233
2.021000E-01	6.200000E-05	PA233
2.053000E-01	3.100000E-05	PA233
2.064000E-01	2.700000E-05	PA233
2.092000E-01	2.300000E-05	PA233
2.158000E-01	2.700000E-05	PA233
2.178000E-01	3.100000E-05	PA233
2.244000E-01	2.300000E-05	PA233
2.252000E-01	9.200000E-05	PA233
2.261000E-01	5.400000E-05	PA233
2.268000E-01	3.100000E-05	PA233
2.285700E-01	4.200000E-05	PA233
2.321000E-01	2.700000E-05	PA233
2.350000E-01	1.200000E-05	PA233
2.360000E-01	2.300000E-05	PA233
2.385000E-01	5.400000E-05	PA233
2.398000E-01	3.100000E-05	PA233
2.423000E-01	5.400000E-05	PA233
2.434000E-01	2.300000E-05	PA233
2.446000E-01	2.700000E-05	PA233
2.483800E-01	6.090000E-04	PA233
2.496000E-01	3.100000E-05	PA233
2.504000E-01	3.100000E-05	PA233
2.523000E-01	3.900000E-05	PA233
2.584500E-01	2.740000E-04	PA233
2.614000E-01	7.800000E-05	PA233
2.644000E-01	3.500000E-05	PA233
2.681000E-01	3.100000E-05	PA233
2.693000E-01	3.100000E-05	PA233
2.715550E-01	3.230000E-03	PA233
2.728000E-01	3.900000E-05	PA233
2.806100E-01	1.100000E-04	PA233
2.884200E-01	1.600000E-04	PA233
2.901000E-01	3.500000E-05	PA233
2.988100E-01	8.799999E-04	PA233
3.001290E-01	6.630000E-02	PA233
3.019900E-01	1.000000E-04	PA233
3.040000E-01	4.600000E-05	PA233
3.054000E-01	5.000000E-05	PA233
3.119040E-01	3.850000E-01	PA233
3.135000E-01	1.390000E-04	PA233
3.176000E-01	2.300000E-05	PA233
3.207300E-01	5.100000E-05	PA233
3.305000E-01	2.300000E-05	PA233
3.359000E-01	2.700000E-05	PA233
3.404760E-01	4.450000E-02	PA233
3.445000E-01	1.500000E-05	PA233
3.518000E-01	4.600000E-05	PA233
3.639000E-01	3.500000E-05	PA233
3.740000E-01	7.300000E-05	PA233
3.754040E-01	6.790000E-03	PA233
3.802800E-01	3.700000E-05	PA233
3.868000E-01	3.100000E-05	PA233
3.933000E-01	5.000000E-05	PA233
3.984920E-01	1.391000E-02	PA233

4.005000E-01 3.100000E-05 PA233
4.029000E-01 2.300000E-05 PA233
4.045000E-01 3.500000E-05 PA233
4.100000E-01 6.899999E-05 PA233
4.143000E-01 5.400000E-05 PA233
4.157640E-01 1.730000E-02 PA233
4.270000E-01 1.900000E-05 PA233
4.328000E-01 8.000000E-06 PA233
4.351000E-01 1.200000E-05 PA233
4.411000E-01 1.900000E-05 PA233
4.542000E-01 2.400000E-05 PA233
4.559600E-01 1.100000E-05 PA233
4.636000E-01 8.000000E-06 PA233
4.711000E-01 1.200000E-05 PA233
4.738000E-01 1.900000E-05 PA233
4.756000E-01 1.900000E-05 PA233
4.780000E-01 1.200000E-05 PA233
4.969000E-01 1.200000E-05 PA233
5.037000E-01 1.200000E-05 PA233
5.063000E-01 1.200000E-05 PA233

8
1.230000E-02 7.099999E-02 TH232
6.381001E-02 2.630000E-03 TH232
8.543100E-02 1.700000E-05 TH232
8.847100E-02 2.800000E-05 TH232
9.943201E-02 3.400000E-06 TH232
1.001300E-01 6.400000E-06 TH232
1.024980E-01 2.400000E-06 TH232
1.408800E-01 2.100000E-04 TH232

75
1.270000E-02 3.150000E-01 PA231
1.650000E-02 3.000000E-03 PA231
1.900000E-02 3.800000E-03 PA231
2.360000E-02 4.700000E-05 PA231
2.450000E-02 5.100000E-05 PA231
2.551000E-02 1.170000E-03 PA231
2.736000E-02 1.030000E-01 PA231
2.996000E-02 1.090000E-03 PA231
3.583000E-02 1.620000E-04 PA231
3.819000E-02 1.600000E-03 PA231
4.415000E-02 6.499999E-04 PA231
4.635000E-02 2.230000E-03 PA231
5.090000E-02 1.550000E-05 PA231
5.273000E-02 8.500000E-04 PA231
5.460000E-02 7.700000E-04 PA231
5.719000E-02 2.990000E-04 PA231
6.050000E-02 6.500000E-05 PA231
6.365000E-02 5.000000E-04 PA231
7.190000E-02 1.900000E-05 PA231
7.272001E-02 3.700000E-05 PA231
7.415000E-02 2.400000E-04 PA231
7.734000E-02 7.300000E-04 PA231
8.767501E-02 7.500000E-03 PA231
9.088600E-02 1.220000E-02 PA231
9.684000E-02 9.500000E-04 PA231
1.008400E-01 3.000000E-04 PA231
1.021010E-01 1.500000E-03 PA231
1.026000E-01 7.000000E-05 PA231
1.028410E-01 2.860000E-03 PA231
1.057380E-01 1.090000E-03 PA231
1.245800E-01 4.800000E-05 PA231

1.443900E-01 1.190000E-04 PA231
1.988900E-01 4.900000E-05 PA231
2.300000E-01 1.700000E-05 PA231
2.430800E-01 4.800000E-04 PA231
2.456000E-01 7.800000E-05 PA231
2.460400E-01 1.110000E-04 PA231
2.557700E-01 1.120000E-03 PA231
2.584400E-01 2.400000E-05 PA231
2.601900E-01 1.880000E-03 PA231
2.731400E-01 6.000000E-04 PA231
2.773200E-01 6.900000E-04 PA231
2.836900E-01 1.700000E-02 PA231
3.000700E-01 2.470000E-02 PA231
3.026500E-01 2.880000E-02 PA231
3.129200E-01 1.020000E-03 PA231
3.271300E-01 3.800000E-04 PA231
3.300600E-01 1.400000E-02 PA231
3.407400E-01 1.810000E-03 PA231
3.515100E-01 7.300000E-05 PA231
3.544600E-01 1.000000E-03 PA231
3.571200E-01 1.750000E-03 PA231
3.593000E-01 8.999999E-05 PA231
3.638400E-01 7.800000E-05 PA231
3.793000E-01 5.000000E-04 PA231
3.847000E-01 3.700000E-05 PA231
3.870000E-01 4.900000E-06 PA231
3.916000E-01 7.800000E-05 PA231
3.955000E-01 2.200000E-05 PA231
3.981400E-01 8.800000E-05 PA231
4.078100E-01 3.600000E-04 PA231
4.103000E-01 3.200000E-05 PA231
4.270000E-01 7.000000E-06 PA231
4.350500E-01 3.110000E-05 PA231
4.380100E-01 4.600000E-05 PA231
4.868300E-01 2.000000E-05 PA231
4.910000E-01 5.300000E-06 PA231
5.014000E-01 8.999999E-06 PA231
5.097001E-01 1.200000E-05 PA231
5.161000E-01 1.000000E-05 PA231
5.358000E-01 6.000000E-06 PA231
5.467001E-01 8.700000E-06 PA231
5.718000E-01 5.300000E-06 PA231
5.830000E-01 4.400000E-05 PA231
6.090000E-01 7.300000E-05 PA231

55
9.200000E-03 3.300000E-04 TH231
1.025000E-02 5.020000E-04 TH231
1.330000E-02 6.000000E-01 TH231
1.720000E-02 2.300000E-03 TH231
1.910000E-02 2.400000E-03 TH231
2.564000E-02 1.410000E-01 TH231
4.286000E-02 5.900000E-04 TH231
4.408000E-02 7.000000E-06 TH231
5.857000E-02 4.620000E-03 TH231
6.386001E-02 2.300000E-04 TH231
6.850000E-02 5.800000E-05 TH231
7.275100E-02 2.520000E-03 TH231
7.769001E-02 4.200000E-05 TH231
8.122800E-02 9.000000E-03 TH231
8.208700E-02 4.200000E-03 TH231
8.421400E-02 6.600000E-02 TH231

8.995000E-02	1.000000E-02	TH231
9.228200E-02	3.700000E-03	TH231
9.302000E-02	4.700000E-04	TH231
9.586301E-02	6.000000E-03	TH231
9.927800E-02	1.310000E-03	TH231
1.022700E-01	4.360000E-03	TH231
1.058100E-01	7.800000E-05	TH231
1.066100E-01	1.760000E-04	TH231
1.075950E-01	7.500000E-04	TH231
1.084220E-01	1.430000E-03	TH231
1.114860E-01	5.500000E-04	TH231
1.156300E-01	1.060000E-05	TH231
1.168200E-01	2.220000E-04	TH231
1.249140E-01	5.800000E-04	TH231
1.340300E-01	2.500000E-04	TH231
1.356640E-01	7.900000E-04	TH231
1.367500E-01	4.400000E-05	TH231
1.405400E-01	7.300000E-06	TH231
1.450600E-01	5.700000E-05	TH231
1.459400E-01	3.170000E-04	TH231
1.631010E-01	1.540000E-03	TH231
1.650000E-01	3.430000E-05	TH231
1.696600E-01	1.320000E-05	TH231
1.741500E-01	1.780000E-04	TH231
1.835000E-01	3.300000E-04	TH231
1.887600E-01	3.300000E-05	TH231
2.179400E-01	3.960000E-04	TH231
2.360100E-01	9.200000E-05	TH231
2.402700E-01	2.900000E-06	TH231
2.425000E-01	8.300000E-06	TH231
2.496000E-01	7.900000E-06	TH231
2.504500E-01	6.600000E-06	TH231
2.676200E-01	1.250000E-05	TH231
2.741000E-01	3.400000E-07	TH231
3.087800E-01	3.600000E-06	TH231
3.110000E-01	3.100000E-05	TH231
3.178700E-01	9.600000E-07	TH231
3.201500E-01	1.500000E-06	TH231
3.518000E-01	6.600000E-07	TH231

18

1.230000E-02	7.700000E-02	TH230
6.767200E-02	3.800000E-03	TH230
8.543100E-02	4.200000E-05	TH230
8.847100E-02	6.800000E-05	TH230
9.943201E-02	8.300000E-06	TH230
1.001300E-01	1.590000E-05	TH230
1.024980E-01	6.000000E-06	TH230
1.100000E-01	6.000000E-07	TH230
1.248000E-01	2.800000E-09	TH230
1.438720E-01	4.899999E-04	TH230
1.860530E-01	8.800000E-05	TH230
2.051000E-01	5.200000E-08	TH230
2.350000E-01	8.400000E-08	TH230
2.537290E-01	1.110000E-04	TH230
2.538000E-01	8.500000E-06	TH230
5.518000E-01	5.500000E-09	TH230
5.705000E-01	3.300000E-08	TH230
6.200000E-01	8.000000E-09	TH230

82

1.230000E-02	7.300000E-01	TH229
1.736000E-02	1.800000E-03	TH229
2.360000E-02	1.230000E-05	TH229
2.539000E-02	8.000001E-05	TH229
3.030000E-02	2.100000E-04	TH229
3.110000E-02	8.400000E-03	TH229
3.150000E-02	1.190000E-02	TH229
3.157000E-02	6.800000E-04	TH229
3.780000E-02	3.300000E-05	TH229
4.230000E-02	8.200000E-04	TH229
4.282000E-02	1.640000E-03	TH229
4.399000E-02	6.600000E-03	TH229
4.975000E-02	2.150000E-04	TH229
5.099000E-02	1.700000E-04	TH229
5.375000E-02	1.100000E-04	TH229
5.511000E-02	2.700000E-05	TH229
5.651800E-02	2.870000E-03	TH229
6.370001E-02	5.100000E-05	TH229
6.809000E-02	6.900000E-04	TH229
6.883001E-02	1.360000E-03	TH229
7.509000E-02	6.100000E-03	TH229
7.763000E-02	4.500000E-04	TH229
7.830001E-02	8.200000E-05	TH229
8.543100E-02	1.470000E-01	TH229
8.625001E-02	1.330000E-02	TH229
8.640000E-02	2.570000E-02	TH229
8.847100E-02	2.400000E-01	TH229
9.473000E-02	2.670000E-03	TH229
9.492000E-02	1.300000E-04	TH229
9.943201E-02	2.930000E-02	TH229
1.001300E-01	5.610000E-02	TH229
1.011000E-01	1.800000E-04	TH229
1.024980E-01	2.120000E-02	TH229
1.046000E-01	8.999999E-05	TH229
1.071080E-01	8.100000E-03	TH229
1.092000E-01	4.400000E-04	TH229
1.103320E-01	1.240000E-03	TH229
1.153000E-01	2.800000E-04	TH229
1.159800E-01	1.700000E-04	TH229
1.179900E-01	1.300000E-04	TH229
1.199800E-01	5.100000E-04	TH229
1.231930E-01	1.510000E-03	TH229
1.245500E-01	6.900000E-03	TH229
1.246500E-01	7.400000E-03	TH229
1.264000E-01	2.100000E-04	TH229
1.265000E-01	1.100000E-04	TH229
1.319260E-01	3.360000E-03	TH229
1.342000E-01	1.050000E-04	TH229
1.369900E-01	1.180000E-02	TH229
1.420000E-01	1.100000E-04	TH229
1.429620E-01	4.040000E-03	TH229
1.476400E-01	2.050000E-03	TH229
1.481500E-01	8.800000E-03	TH229
1.500400E-01	3.000000E-04	TH229
1.516000E-01	2.600000E-04	TH229
1.543360E-01	7.699999E-03	TH229
1.564090E-01	1.190000E-02	TH229
1.584200E-01	4.800000E-04	TH229
1.633400E-01	2.100000E-04	TH229
1.669760E-01	2.050000E-03	TH229
1.674500E-01	5.100000E-04	TH229
1.717500E-01	2.100000E-04	TH229
1.729260E-01	1.130000E-03	TH229
1.742200E-01	1.800000E-04	TH229
1.748200E-01	1.500000E-04	TH229

1.797570E-01	1.970000E-03	TH229
1.839280E-01	1.420000E-03	TH229
1.935090E-01	4.410000E-02	TH229
1.943000E-01	3.100000E-04	TH229
2.008070E-01	6.900000E-04	TH229
2.046900E-01	6.000000E-03	TH229
2.101500E-01	1.900000E-03	TH229
2.108530E-01	2.800000E-02	TH229
2.151000E-01	1.370000E-03	TH229
2.181540E-01	1.850000E-03	TH229
2.212200E-01	2.300000E-04	TH229
2.251490E-01	7.200000E-04	TH229
2.362490E-01	1.740000E-03	TH229
2.424000E-01	9.399999E-04	TH229
2.524300E-01	9.500000E-04	TH229
2.590800E-01	3.400000E-04	TH229
2.962000E-01	1.200000E-04	TH229

20

1.230000E-02	8.800000E-02	TH228
7.440001E-02	4.000000E-06	TH228
8.437300E-02	1.220000E-02	TH228
8.543100E-02	1.850000E-04	TH228
8.847100E-02	3.020000E-04	TH228
9.943201E-02	3.690000E-05	TH228
1.001300E-01	7.060000E-05	TH228
1.024980E-01	2.670000E-05	TH228
1.316130E-01	1.305000E-03	TH228
1.420000E-01	1.300000E-08	TH228
1.664100E-01	1.036000E-03	TH228
1.822000E-01	5.200000E-08	TH228
2.059300E-01	1.960000E-04	TH228
2.159830E-01	2.540000E-03	TH228
2.285000E-01	1.800000E-07	TH228
7.005000E-01	3.100000E-08	TH228
7.422001E-01	1.500000E-08	TH228
8.320000E-01	1.460000E-07	TH228
9.081000E-01	1.710000E-08	TH228
9.929001E-01	1.460000E-08	TH228

239

1.300000E-02	3.370000E-01	AC228
1.840000E-02	1.400000E-04	AC228
4.246000E-02	8.999999E-05	AC228
5.776600E-02	4.700000E-03	AC228
7.734000E-02	2.600000E-04	AC228
8.995701E-02	1.900000E-02	AC228
9.335000E-02	3.100000E-02	AC228
9.950901E-02	1.260000E-02	AC228
1.004100E-01	9.300000E-04	AC228
1.048190E-01	3.800000E-03	AC228
1.056040E-01	7.200000E-03	AC228
1.085820E-01	2.800000E-03	AC228
1.145600E-01	9.800000E-05	AC228
1.290650E-01	2.420000E-02	AC228
1.355400E-01	1.800000E-04	AC228
1.410200E-01	5.000000E-04	AC228
1.458490E-01	1.580000E-03	AC228
1.539770E-01	7.220000E-03	AC228
1.686500E-01	1.300000E-04	AC228
1.739640E-01	3.500000E-04	AC228
1.845400E-01	7.000000E-04	AC228
1.913530E-01	1.230000E-03	AC228

1.994070E-01	3.150000E-03	AC228
2.040260E-01	1.120000E-03	AC228
2.092530E-01	3.890000E-02	AC228
2.148500E-01	2.900000E-04	AC228
2.238500E-01	5.400000E-04	AC228
2.314200E-01	2.500000E-04	AC228
2.575200E-01	3.000000E-04	AC228
2.635800E-01	4.000000E-04	AC228
2.702450E-01	3.460000E-02	AC228
2.789500E-01	1.910000E-03	AC228
2.820000E-01	7.200000E-04	AC228
3.216460E-01	2.260000E-03	AC228
3.260400E-01	3.300000E-04	AC228
3.274400E-01	1.200000E-03	AC228
3.280000E-01	2.950000E-02	AC228
3.323700E-01	4.000000E-03	AC228
3.383200E-01	1.127000E-01	AC228
3.409600E-01	3.690000E-03	AC228
3.569400E-01	1.700000E-04	AC228
3.725700E-01	6.699999E-05	AC228
3.779900E-01	2.500000E-04	AC228
3.846300E-01	6.699999E-05	AC228
3.891200E-01	1.030000E-04	AC228
3.979400E-01	2.700000E-04	AC228
3.996200E-01	2.900000E-04	AC228
4.094620E-01	1.920000E-02	AC228
4.163000E-01	1.320000E-04	AC228
4.194200E-01	2.100000E-04	AC228
4.404400E-01	1.210000E-03	AC228
4.491500E-01	4.800000E-04	AC228
4.524700E-01	1.500000E-04	AC228
4.571700E-01	1.500000E-04	AC228
4.630040E-01	4.400000E-02	AC228
4.702500E-01	1.300000E-04	AC228
4.717600E-01	3.300000E-04	AC228
4.747500E-01	2.200000E-04	AC228
4.783300E-01	2.090000E-03	AC228
4.809400E-01	2.300000E-04	AC228
4.903300E-01	1.110000E-04	AC228
4.923700E-01	2.350000E-04	AC228
4.974900E-01	5.900000E-05	AC228
5.038230E-01	1.820000E-03	AC228
5.089591E-01	4.500000E-03	AC228
5.150600E-01	4.899999E-04	AC228
5.201510E-01	6.700000E-04	AC228
5.231310E-01	1.030000E-03	AC228
5.407600E-01	2.600000E-04	AC228
5.464700E-01	2.010000E-03	AC228
5.487300E-01	2.300000E-04	AC228
5.551201E-01	4.600000E-04	AC228
5.625000E-01	8.700000E-03	AC228
5.709100E-01	1.820000E-03	AC228
5.721400E-01	1.500000E-03	AC228
5.834100E-01	1.110000E-03	AC228
5.904000E-01	1.700000E-04	AC228
6.106400E-01	2.300000E-04	AC228
6.162200E-01	8.000000E-04	AC228
6.203800E-01	8.000000E-04	AC228
6.232700E-01	1.100000E-04	AC228
6.272300E-01	1.400000E-04	AC228
6.294001E-01	4.500000E-04	AC228
6.403400E-01	5.400000E-04	AC228
6.488401E-01	8.000000E-04	AC228

6.515101E-01	9.000000E-04	AC228
6.601000E-01	5.000000E-05	AC228
6.638200E-01	2.800000E-04	AC228
6.664500E-01	6.199999E-04	AC228
6.720001E-01	2.600000E-04	AC228
6.741600E-01	5.000000E-04	AC228
6.747500E-01	5.000000E-04	AC228
6.771100E-01	6.199999E-04	AC228
6.840000E-01	1.900000E-04	AC228
6.881000E-01	1.340000E-03	AC228
6.925001E-01	5.600000E-05	AC228
6.990800E-01	3.700000E-04	AC228
7.017471E-01	1.730000E-03	AC228
7.074100E-01	1.550000E-03	AC228
7.184800E-01	1.900000E-04	AC228
7.268630E-01	6.200000E-03	AC228
7.377200E-01	3.700000E-04	AC228
7.553151E-01	1.000000E-02	AC228
7.700400E-01	6.300000E-05	AC228
7.722911E-01	1.490000E-02	AC228
7.741000E-01	6.000000E-04	AC228
7.765600E-01	1.900000E-04	AC228
7.782300E-01	2.200000E-04	AC228
7.821420E-01	4.850000E-03	AC228
7.914900E-01	2.300000E-04	AC228
7.928000E-01	8.000000E-04	AC228
7.949471E-01	4.250000E-02	AC228
8.137701E-01	7.000000E-05	AC228
8.167101E-01	3.000000E-04	AC228
8.249341E-01	5.000000E-04	AC228
8.304861E-01	5.400000E-03	AC228
8.357100E-01	1.610000E-02	AC228
8.403770E-01	9.100000E-03	AC228
8.531700E-01	3.100000E-05	AC228
8.704601E-01	4.400000E-04	AC228
8.731700E-01	3.100000E-04	AC228
8.744400E-01	4.700000E-04	AC228
8.774601E-01	1.400000E-04	AC228
8.807601E-01	6.200000E-05	AC228
8.873301E-01	2.700000E-04	AC228
9.012300E-01	1.600000E-04	AC228
9.042001E-01	7.699999E-03	AC228
9.112040E-01	2.580000E-01	AC228
9.189700E-01	2.700000E-04	AC228
9.219800E-01	2.940000E-04	AC228
9.240301E-01	7.500000E-05	AC228
9.309300E-01	2.480000E-04	AC228
9.398701E-01	8.999999E-05	AC228
9.441960E-01	9.500000E-04	AC228
9.479820E-01	1.060000E-03	AC228
9.586101E-01	2.800000E-03	AC228
9.647660E-01	4.990000E-02	AC228
9.689711E-01	1.580000E-01	AC228
9.759601E-01	5.000000E-04	AC228
9.794800E-01	2.600000E-04	AC228
9.877101E-01	7.700000E-04	AC228
9.886301E-01	7.700000E-04	AC228
1.000690E+00	5.000000E-05	AC228
1.013580E+00	4.600000E-05	AC228
1.016440E+00	3.800000E-04	AC228
1.017920E+00	5.700000E-05	AC228
1.019860E+00	2.100000E-04	AC228
1.033248E+00	2.010000E-03	AC228
1.039650E+00	4.400000E-04	AC228
1.040920E+00	4.400000E-04	AC228
1.053090E+00	1.300000E-04	AC228
1.054110E+00	1.800000E-04	AC228
1.062550E+00	1.000000E-04	AC228
1.065180E+00	1.320000E-03	AC228
1.074710E+00	1.000000E-04	AC228
1.088180E+00	5.900000E-05	AC228
1.095679E+00	1.290000E-03	AC228
1.103410E+00	1.500000E-04	AC228
1.110610E+00	3.040000E-03	AC228
1.117630E+00	5.400000E-04	AC228
1.135240E+00	9.800000E-05	AC228
1.142850E+00	1.030000E-04	AC228
1.148120E+00	5.900000E-05	AC228
1.153520E+00	1.390000E-03	AC228
1.157140E+00	7.000000E-05	AC228
1.164500E+00	6.499999E-04	AC228
1.175310E+00	2.400000E-04	AC228
1.190810E+00	6.200000E-05	AC228
1.217030E+00	2.100000E-04	AC228
1.229400E+00	7.500000E-05	AC228
1.245050E+00	9.500000E-04	AC228
1.247080E+00	5.000000E-03	AC228
1.250040E+00	6.199999E-04	AC228
1.276690E+00	1.400000E-04	AC228
1.286270E+00	5.000000E-04	AC228
1.287680E+00	8.000000E-04	AC228
1.309710E+00	1.900000E-04	AC228
1.315340E+00	1.500000E-04	AC228
1.344590E+00	8.999999E-05	AC228
1.347500E+00	1.500000E-04	AC228
1.357780E+00	2.000000E-04	AC228
1.365700E+00	1.400000E-04	AC228
1.374190E+00	1.400000E-04	AC228
1.401490E+00	1.200000E-04	AC228
1.415660E+00	2.100000E-04	AC228
1.430950E+00	3.500000E-04	AC228
1.451400E+00	1.060000E-04	AC228
1.459138E+00	8.300000E-03	AC228
1.469710E+00	2.000000E-04	AC228
1.495910E+00	8.600000E-03	AC228
1.501570E+00	4.600000E-03	AC228
1.537890E+00	4.700000E-04	AC228
1.548650E+00	3.800000E-04	AC228
1.557110E+00	1.780000E-03	AC228
1.559850E+00	2.000000E-04	AC228
1.571520E+00	5.700000E-05	AC228
1.573260E+00	3.300000E-04	AC228
1.580530E+00	6.000000E-03	AC228
1.588200E+00	3.220000E-02	AC228
1.609410E+00	7.700000E-05	AC228
1.625060E+00	2.550000E-03	AC228
1.630627E+00	1.510000E-02	AC228
1.638281E+00	4.700000E-03	AC228
1.666523E+00	1.780000E-03	AC228
1.677670E+00	5.400000E-04	AC228
1.686090E+00	9.500000E-04	AC228
1.700590E+00	1.010000E-04	AC228
1.702430E+00	4.800000E-04	AC228
1.706190E+00	8.500000E-05	AC228
1.713470E+00	5.400000E-05	AC228
1.724210E+00	2.900000E-04	AC228

1.738220E+00	1.800000E-04	AC228
1.740400E+00	1.100000E-04	AC228
1.742000E+00	8.000001E-05	AC228
1.750540E+00	8.000001E-05	AC228
1.758110E+00	3.500000E-04	AC228
1.772200E+00	1.800000E-05	AC228
1.795100E+00	2.100000E-05	AC228
1.797500E+00	2.100000E-05	AC228
1.800860E+00	4.400000E-05	AC228
1.823220E+00	4.400000E-04	AC228
1.826700E+00	2.100000E-05	AC228
1.835430E+00	3.800000E-04	AC228
1.842130E+00	4.200000E-04	AC228
1.850130E+00	4.400000E-05	AC228
1.870830E+00	2.430000E-04	AC228
1.879600E+00	1.300000E-05	AC228
1.887100E+00	9.000000E-04	AC228
1.900070E+00	2.800000E-05	AC228
1.907180E+00	1.190000E-04	AC228
1.929780E+00	1.990000E-04	AC228
1.936300E+00	2.100000E-05	AC228
1.952330E+00	5.900000E-04	AC228
1.955900E+00	8.000000E-06	AC228
1.958400E+00	1.500000E-05	AC228
1.965240E+00	2.040000E-04	AC228
1.971900E+00	3.600000E-05	AC228
1.979300E+00	1.800000E-05	AC228
2.029400E+00	1.800000E-05	AC228

5

6.280000E-03	1.410000E-08	RA228
6.670000E-03	3.100000E-07	RA228
1.275000E-02	3.100000E-03	RA228
1.352000E-02	1.550000E-02	RA228
2.640000E-02	1.400000E-04	RA228

206

6.500000E-03	1.200000E-03	TH227
8.150000E-03	1.000000E-04	TH227
2.025000E-02	3.200000E-03	TH227
2.960000E-02	7.000000E-05	TH227
2.986000E-02	1.030000E-03	TH227
3.158000E-02	9.300000E-04	TH227
3.339000E-02	1.000000E-04	TH227
4.020000E-02	2.100000E-04	TH227
4.193000E-02	3.900000E-04	TH227
4.377000E-02	2.900000E-03	TH227
4.380000E-02	8.000000E-04	TH227
4.422000E-02	7.200000E-04	TH227
4.440000E-02	2.200000E-04	TH227
4.830000E-02	1.900000E-04	TH227
4.982000E-02	5.800000E-03	TH227
5.013000E-02	1.140000E-01	TH227
5.085000E-02	2.100000E-04	TH227
5.419000E-02	9.570000E-05	TH227
5.600000E-02	6.699999E-05	TH227
5.642000E-02	1.200000E-04	TH227
6.144100E-02	1.220000E-03	TH227
6.245000E-02	5.400000E-03	TH227
6.268001E-02	1.000000E-04	TH227
6.435000E-02	3.500000E-04	TH227
6.874000E-02	1.580000E-03	TH227
6.980000E-02	1.400000E-04	TH227

7.285000E-02	3.000000E-04	TH227
7.363000E-02	1.900000E-04	TH227
7.501001E-02	3.700000E-04	TH227
7.969001E-02	2.640000E-02	TH227
8.543100E-02	1.810000E-02	TH227
8.847100E-02	2.960000E-02	TH227
8.960000E-02	5.200000E-05	TH227
9.388001E-02	2.050000E-02	TH227
9.497000E-02	6.600000E-04	TH227
9.603000E-02	9.500000E-04	TH227
9.943201E-02	3.620000E-03	TH227
9.958000E-02	3.500000E-04	TH227
9.960000E-02	1.750000E-04	TH227
1.001300E-01	6.900000E-03	TH227
1.002700E-01	1.140000E-03	TH227
1.024980E-01	2.620000E-03	TH227
1.025000E-01	1.570000E-05	TH227
1.077600E-01	2.000000E-04	TH227
1.092000E-01	7.200000E-05	TH227
1.106500E-01	4.000000E-05	TH227
1.131100E-01	9.500000E-03	TH227
1.172000E-01	2.700000E-03	TH227
1.235800E-01	1.900000E-04	TH227
1.244400E-01	6.000000E-05	TH227
1.346000E-01	4.600000E-04	TH227
1.414200E-01	3.200000E-03	TH227
1.501400E-01	1.500000E-04	TH227
1.621900E-01	1.000000E-04	TH227
1.683600E-01	2.000000E-04	TH227
1.699500E-01	8.000001E-05	TH227
1.734500E-01	2.400000E-04	TH227
1.758000E-01	2.800000E-04	TH227
1.846500E-01	4.899999E-04	TH227
1.975600E-01	1.700000E-04	TH227
2.005000E-01	1.700000E-04	TH227
2.016400E-01	3.200000E-04	TH227
2.041400E-01	3.100000E-03	TH227
2.049800E-01	2.200000E-03	TH227
2.060800E-01	3.400000E-03	TH227
2.106200E-01	1.700000E-02	TH227
2.127000E-01	1.330000E-03	TH227
2.189000E-01	2.980000E-03	TH227
2.255000E-01	1.200000E-04	TH227
2.299000E-01	5.200000E-05	TH227
2.347600E-01	6.100000E-03	TH227
2.359600E-01	1.750000E-01	TH227
2.461200E-01	1.660000E-04	TH227
2.496000E-01	1.000000E-04	TH227
2.501500E-01	1.210000E-04	TH227
2.502700E-01	6.100000E-03	TH227
2.525000E-01	1.510000E-03	TH227
2.546300E-01	9.600000E-03	TH227
2.562300E-01	9.500000E-02	TH227
2.628700E-01	1.450000E-03	TH227
2.670500E-01	1.400000E-04	TH227
2.678600E-01	1.000000E-04	TH227
2.705600E-01	3.900000E-04	TH227
2.729100E-01	6.900000E-03	TH227
2.798000E-01	7.400000E-04	TH227
2.807000E-01	3.500000E-05	TH227
2.814200E-01	4.820000E-03	TH227
2.842400E-01	5.400000E-04	TH227
2.855200E-01	6.000000E-04	TH227

2.860900E-01	2.400000E-02	TH227
2.895900E-01	2.600000E-02	TH227
2.897700E-01	2.600000E-04	TH227
2.924100E-01	8.900000E-04	TH227
2.965000E-01	6.000000E-03	TH227
2.999800E-01	3.000000E-02	TH227
3.005000E-01	1.900000E-04	TH227
3.045000E-01	1.560000E-02	TH227
3.084000E-01	2.300000E-04	TH227
3.126900E-01	7.000000E-03	TH227
3.148500E-01	1.320000E-02	TH227
3.184600E-01	8.999999E-05	TH227
3.192400E-01	4.400000E-04	TH227
3.248800E-01	1.400000E-04	TH227
3.259900E-01	8.999999E-05	TH227
3.298500E-01	4.000000E-02	TH227
3.343700E-01	1.540000E-02	TH227
3.397600E-01	5.200000E-05	TH227
3.425500E-01	4.700000E-03	TH227
3.464500E-01	1.630000E-04	TH227
3.505400E-01	1.500000E-03	TH227
3.526100E-01	1.400000E-04	TH227
3.626300E-01	6.900000E-04	TH227
3.693500E-01	8.399999E-05	TH227
3.709300E-01	5.000000E-05	TH227
3.748000E-01	2.100000E-05	TH227
3.762700E-01	7.000000E-05	TH227
3.822000E-01	8.700000E-05	TH227
3.835100E-01	3.000000E-04	TH227
3.986000E-01	1.900000E-05	TH227
4.022000E-01	1.000000E-04	TH227
4.151100E-01	1.900000E-05	TH227
4.323300E-01	5.600000E-05	TH227
4.480000E-01	1.900000E-06	TH227
4.575000E-01	9.500000E-07	TH227
4.668000E-01	6.700000E-06	TH227
4.800000E-01	4.000000E-06	TH227
4.820000E-01	1.900000E-06	TH227
4.931000E-01	7.300000E-06	TH227
5.075001E-01	8.999999E-06	TH227
5.166000E-01	3.900000E-06	TH227
5.245000E-01	2.600000E-06	TH227
5.346000E-01	1.300000E-06	TH227
5.369000E-01	1.500000E-05	TH227
5.524001E-01	3.100000E-06	TH227
5.561000E-01	5.100000E-06	TH227
5.690000E-01	8.100000E-06	TH227
5.760000E-01	4.000000E-06	TH227
5.790001E-01	6.000000E-06	TH227
5.890000E-01	8.000000E-07	TH227
5.960000E-01	1.400000E-07	TH227
6.077000E-01	2.500000E-06	TH227
6.214001E-01	8.000000E-07	TH227
6.238000E-01	2.300000E-06	TH227
6.323000E-01	1.900000E-06	TH227
6.410000E-01	2.600000E-07	TH227
6.443000E-01	1.200000E-06	TH227
6.628000E-01	8.000000E-07	TH227
6.920000E-01	5.400000E-07	TH227
7.043000E-01	1.090000E-06	TH227
7.072001E-01	5.400000E-07	TH227
7.185000E-01	4.000000E-07	TH227
7.221000E-01	5.100000E-06	TH227

7.235000E-01	3.700000E-06	TH227
7.344000E-01	1.400000E-06	TH227
7.354001E-01	2.300000E-06	TH227
7.384000E-01	9.500000E-07	TH227
7.464001E-01	1.400000E-06	TH227
7.488000E-01	5.600000E-06	TH227
7.541000E-01	6.600000E-06	TH227
7.569001E-01	5.200000E-06	TH227
7.622001E-01	3.700000E-06	TH227
7.663000E-01	4.000000E-06	TH227
7.734001E-01	2.100000E-06	TH227
7.758000E-01	2.100000E-05	TH227
7.810000E-01	4.400000E-06	TH227
7.842001E-01	1.300000E-06	TH227
7.874001E-01	1.000000E-06	TH227
7.926000E-01	5.400000E-07	TH227
7.973000E-01	1.240000E-05	TH227
8.039001E-01	8.999999E-06	TH227
8.086000E-01	1.000000E-06	TH227
8.126000E-01	2.300000E-05	TH227
8.180000E-01	2.300000E-06	TH227
8.181000E-01	2.300000E-06	TH227
8.234001E-01	3.500000E-05	TH227
8.267000E-01	2.300000E-06	TH227
8.285000E-01	2.600000E-06	TH227
8.378000E-01	7.200000E-06	TH227
8.425000E-01	1.210000E-05	TH227
8.467001E-01	2.000000E-06	TH227
8.483000E-01	3.700000E-06	TH227
8.543000E-01	9.500000E-07	TH227
8.573000E-01	8.000000E-07	TH227
8.589001E-01	3.500000E-06	TH227
8.630000E-01	2.600000E-07	TH227
8.673000E-01	4.000000E-06	TH227
8.763000E-01	3.100000E-06	TH227
8.782001E-01	1.900000E-06	TH227
8.910000E-01	2.600000E-07	TH227
8.930001E-01	1.800000E-07	TH227
8.961000E-01	1.500000E-06	TH227
9.086000E-01	3.200000E-05	TH227
9.100000E-01	2.100000E-07	TH227
9.200000E-01	1.600000E-07	TH227
9.270000E-01	9.000000E-08	TH227
9.380000E-01	1.400000E-07	TH227
9.416000E-01	9.600000E-07	TH227
9.587001E-01	8.400000E-07	TH227
9.700000E-01	1.900000E-06	TH227
9.717001E-01	1.400000E-07	TH227
9.900001E-01	4.700000E-07	TH227
9.950001E-01	9.000000E-08	TH227
9.998000E-01	4.000000E-07	TH227
1.015200E+00	2.100000E-07	TH227
1.020000E+00	2.600000E-07	TH227
1.025000E+00	2.100000E-07	TH227

69
1.200000E-02	8.200000E-04	AC227A
1.290000E-02	8.000000E-08	AC227A
2.595000E-02	6.100000E-09	AC227A
3.350000E-02	1.200000E-06	AC227A
3.500000E-02	3.000000E-07	AC227A
3.747000E-02	3.000000E-08	AC227A
4.470000E-02	1.200000E-06	AC227A

5.106000E-02	3.000000E-09	AC227A
5.232000E-02	1.500000E-08	AC227A
5.370000E-02	4.600000E-07	AC227A
5.500000E-02	4.900000E-06	AC227A
5.580000E-02	4.300000E-08	AC227A
5.756000E-02	3.500000E-08	AC227A
5.940000E-02	4.600000E-07	AC227A
6.060000E-02	4.600000E-07	AC227A
6.928000E-02	4.300000E-05	AC227A
7.060000E-02	9.000000E-07	AC227A
7.250001E-02	1.600000E-06	AC227A
7.954001E-02	1.200000E-05	AC227A
8.220000E-02	8.999999E-06	AC227A
8.300000E-02	1.500000E-08	AC227A
8.323101E-02	3.500000E-05	AC227A
8.500000E-02	1.200000E-07	AC227A
8.610000E-02	5.200000E-06	AC227A
8.610500E-02	6.000000E-05	AC227A
8.670000E-02	3.000000E-05	AC227A
8.810000E-02	1.600000E-05	AC227A
8.850000E-02	1.100000E-08	AC227A
9.000000E-02	2.000000E-06	AC227A
9.681500E-02	7.000000E-06	AC227A
9.747400E-02	1.300000E-05	AC227A
9.960000E-02	5.600000E-05	AC227A
1.002140E-01	5.000000E-06	AC227A
1.010000E-01	8.000000E-06	AC227A
1.050000E-01	3.800000E-06	AC227A
1.068500E-01	1.200000E-05	AC227A
1.080000E-01	4.600000E-07	AC227A
1.187000E-01	4.600000E-07	AC227A
1.216000E-01	2.800000E-05	AC227A
1.345000E-01	6.100000E-06	AC227A
1.374000E-01	4.600000E-06	AC227A
1.409000E-01	2.300000E-06	AC227A
1.430000E-01	6.000000E-06	AC227A
1.436500E-01	2.900000E-07	AC227A
1.460000E-01	9.999999E-08	AC227A
1.476100E-01	2.700000E-05	AC227A
1.493000E-01	1.500000E-07	AC227A
1.592000E-01	6.100000E-06	AC227A
1.604900E-01	4.900000E-05	AC227A
1.614000E-01	1.500000E-06	AC227A
1.626000E-01	6.000000E-07	AC227A
1.720000E-01	1.100000E-05	AC227A
1.743000E-01	3.000000E-06	AC227A
1.761000E-01	7.200000E-06	AC227A
2.068000E-01	1.100000E-05	AC227A
2.166000E-01	6.000000E-07	AC227A
2.297000E-01	4.600000E-06	AC227A
2.309000E-01	1.500000E-07	AC227A
2.317900E-01	7.900000E-08	AC227A
2.426000E-01	3.000000E-06	AC227A
2.439000E-01	3.000000E-07	AC227A
2.834000E-01	6.000000E-07	AC227A
3.517000E-01	6.000000E-07	AC227A
4.156000E-01	2.300000E-06	AC227A
4.396000E-01	3.800000E-07	AC227A
4.410000E-01	6.000000E-07	AC227A
4.602000E-01	2.300000E-06	AC227A
5.276000E-01	3.200000E-07	AC227A
5.404000E-01	7.699999E-07	AC227A

11

1.170000E-02	8.000000E-03	RA226
8.106901E-02	1.930000E-03	RA226
8.378701E-02	3.190000E-03	RA226
9.424701E-02	3.850000E-04	RA226
9.486800E-02	7.350000E-04	RA226
9.753000E-02	2.760000E-04	RA226
1.862110E-01	3.590000E-02	RA226
2.622700E-01	5.000000E-05	RA226
4.146000E-01	3.000000E-06	RA226
4.493700E-01	1.900000E-06	RA226
6.006600E-01	4.900000E-06	RA226

113

1.064200E-02	8.500000E-02	AC225
1.200000E-02	1.300000E-01	AC225
2.600000E-02	1.500000E-05	AC225
3.670000E-02	1.550000E-04	AC225
3.850000E-02	9.400000E-05	AC225
4.620000E-02	3.900000E-05	AC225
4.910000E-02	6.600000E-05	AC225
5.780000E-02	3.900000E-05	AC225
6.290001E-02	4.300000E-03	AC225
6.430001E-02	4.100000E-04	AC225
6.987000E-02	4.700000E-05	AC225
7.140000E-02	1.260000E-04	AC225
7.350000E-02	2.500000E-04	AC225
7.390001E-02	2.640000E-03	AC225
7.460000E-02	2.200000E-04	AC225
7.880001E-02	1.070000E-04	AC225
8.323101E-02	7.500000E-03	AC225
8.610500E-02	1.230000E-02	AC225
8.740000E-02	2.260000E-03	AC225
9.490000E-02	8.400000E-04	AC225
9.670001E-02	2.800000E-04	AC225
9.681500E-02	1.500000E-03	AC225
9.747400E-02	2.900000E-03	AC225
9.960000E-02	7.000000E-03	AC225
9.980001E-02	1.000000E-02	AC225
1.002140E-01	1.080000E-03	AC225
1.008000E-01	7.500000E-04	AC225
1.036000E-01	2.300000E-05	AC225
1.084000E-01	2.160000E-03	AC225
1.115000E-01	2.640000E-03	AC225
1.128000E-01	1.800000E-05	AC225
1.140000E-01	7.500000E-06	AC225
1.199000E-01	6.600000E-04	AC225
1.238000E-01	7.200000E-04	AC225
1.248000E-01	2.400000E-04	AC225
1.262000E-01	7.000000E-05	AC225
1.292000E-01	2.200000E-05	AC225
1.336000E-01	1.700000E-04	AC225
1.349000E-01	2.700000E-04	AC225
1.376000E-01	1.900000E-05	AC225
1.396000E-01	1.200000E-05	AC225
1.447000E-01	4.000000E-06	AC225
1.452000E-01	1.260000E-03	AC225
1.501000E-01	6.000000E-03	AC225
1.526000E-01	1.900000E-04	AC225
1.539000E-01	1.820000E-03	AC225
1.573000E-01	3.200000E-03	AC225
1.691000E-01	7.000000E-05	AC225
1.699000E-01	1.200000E-04	AC225

1.707000E-01	1.700000E-04	AC225
1.783000E-01	1.400000E-04	AC225
1.798000E-01	9.400000E-05	AC225
1.861000E-01	1.100000E-04	AC225
1.863000E-01	3.600000E-05	AC225
1.872000E-01	8.899999E-05	AC225
1.880000E-01	4.500000E-03	AC225
1.958000E-01	1.230000E-03	AC225
1.974000E-01	2.300000E-04	AC225
1.979000E-01	3.300000E-04	AC225
1.984000E-01	1.700000E-04	AC225
2.047000E-01	1.100000E-05	AC225
2.169000E-01	2.710000E-03	AC225
2.247000E-01	9.799999E-04	AC225
2.282000E-01	4.000000E-05	AC225
2.360000E-01	1.500000E-05	AC225
2.407000E-01	1.000000E-04	AC225
2.432000E-01	3.000000E-05	AC225
2.496000E-01	1.200000E-04	AC225
2.535000E-01	1.160000E-03	AC225
2.560000E-01	6.000000E-06	AC225
2.793000E-01	2.500000E-04	AC225
2.848000E-01	6.300000E-05	AC225
2.986000E-01	1.800000E-05	AC225
3.174000E-01	1.100000E-06	AC225
3.218000E-01	3.000000E-05	AC225
3.549000E-01	2.300000E-05	AC225
3.566000E-01	2.290000E-06	AC225
3.622000E-01	4.200000E-05	AC225
3.683000E-01	6.000000E-06	AC225
3.750000E-01	1.700000E-05	AC225
4.034000E-01	1.600000E-06	AC225
4.062000E-01	6.699999E-05	AC225
4.179000E-01	4.800000E-05	AC225
4.350000E-01	2.400000E-05	AC225
4.501000E-01	3.200000E-05	AC225
4.524000E-01	8.900000E-04	AC225
4.588000E-01	5.800000E-06	AC225
4.624000E-01	8.000000E-06	AC225
4.695000E-01	2.800000E-05	AC225
4.811000E-01	2.900000E-04	AC225
4.926000E-01	2.200000E-06	AC225
5.125000E-01	5.000000E-06	AC225
5.153000E-01	1.900000E-04	AC225
5.179000E-01	1.500000E-04	AC225
5.221000E-01	1.800000E-05	AC225
5.261000E-01	3.300000E-04	AC225
5.297000E-01	7.100000E-05	AC225
5.312001E-01	4.000000E-05	AC225
5.458000E-01	4.600000E-06	AC225
5.520000E-01	5.600000E-05	AC225
5.656000E-01	1.900000E-06	AC225
5.683000E-01	1.300000E-05	AC225
5.710000E-01	3.200000E-05	AC225
5.914000E-01	7.000000E-06	AC225
5.946000E-01	2.800000E-05	AC225
6.010000E-01	3.700000E-05	AC225
6.035000E-01	1.600000E-05	AC225
6.299000E-01	2.600000E-06	AC225
6.371000E-01	1.000000E-06	AC225
6.495000E-01	1.200000E-05	AC225
6.804001E-01	6.100000E-06	AC225
7.806000E-01	5.000000E-07	AC225

8.242000E-01	4.000000E-07	AC225

2
1.270000E-02	1.380000E-01	RA225
4.000000E-02	3.000000E-01	RA225

5
6.280000E-03	1.410000E-08	RA224
6.670000E-03	3.100000E-07	RA224
1.275000E-02	3.100000E-03	RA224
1.352000E-02	1.550000E-02	RA224
2.640000E-02	1.400000E-04	RA224

83
1.000000E-02	1.390000E-04	RA223
1.170000E-02	2.290000E-01	RA223
1.440000E-02	1.670000E-04	RA223
3.198000E-02	9.700000E-07	RA223
6.950001E-02	7.000000E-05	RA223
7.090001E-02	3.500000E-05	RA223
8.106901E-02	1.500000E-01	RA223
8.378701E-02	2.470000E-01	RA223
9.424701E-02	2.980000E-02	RA223
9.486800E-02	5.690000E-02	RA223
9.753000E-02	2.140000E-02	RA223
1.022000E-01	8.000000E-06	RA223
1.032000E-01	6.000000E-05	RA223
1.042300E-01	1.900000E-04	RA223
1.067800E-01	2.360000E-04	RA223
1.085000E-01	6.000000E-05	RA223
1.108560E-01	5.800000E-04	RA223
1.147000E-01	1.000000E-04	RA223
1.223190E-01	1.209000E-02	RA223
1.316000E-01	6.000000E-05	RA223
1.383000E-01	1.700000E-05	RA223
1.442350E-01	3.270000E-02	RA223
1.472000E-01	6.000000E-05	RA223
1.542080E-01	5.700000E-02	RA223
1.586350E-01	6.950000E-03	RA223
1.658000E-01	6.000000E-05	RA223
1.756500E-01	1.900000E-04	RA223
1.773000E-01	4.700000E-04	RA223
1.795400E-01	1.530000E-03	RA223
1.993000E-01	2.800000E-05	RA223
2.213200E-01	3.600000E-04	RA223
2.472000E-01	1.000000E-04	RA223
2.493000E-01	3.900000E-04	RA223
2.516000E-01	4.200000E-04	RA223
2.552000E-01	5.300000E-04	RA223
2.557000E-01	6.000000E-05	RA223
2.604000E-01	7.000000E-05	RA223
2.694630E-01	1.390000E-01	RA223
2.703000E-01	7.000000E-06	RA223
2.860000E-01	1.100000E-05	RA223
2.881800E-01	1.600000E-03	RA223
3.238710E-01	3.990000E-02	RA223
3.283800E-01	2.090000E-03	RA223
3.340100E-01	1.010000E-03	RA223
3.382820E-01	2.840000E-02	RA223
3.428700E-01	2.220000E-03	RA223
3.555000E-01	4.200000E-05	RA223
3.557000E-01	2.800000E-05	RA223
3.620520E-01	4.600000E-04	RA223

3.629000E-01 1.500000E-04 RA223
3.685600E-01 8.000001E-05 RA223
3.716760E-01 4.870000E-03 RA223
3.729000E-01 5.000000E-04 RA223
3.761000E-01 1.300000E-04 RA223
3.828000E-01 1.400000E-04 RA223
3.877000E-01 1.500000E-04 RA223
3.901000E-01 7.000000E-05 RA223
4.306000E-01 1.900000E-04 RA223
4.321200E-01 3.500000E-04 RA223
4.450330E-01 1.290000E-02 RA223
4.875000E-01 1.110000E-04 RA223
5.000000E-01 1.400000E-05 RA223
5.100001E-01 4.000000E-06 RA223
5.276111E-01 7.100000E-04 RA223
5.376000E-01 2.090000E-05 RA223
5.420001E-01 1.400000E-05 RA223
5.458000E-01 1.100000E-05 RA223
5.741000E-01 1.100000E-05 RA223
5.796000E-01 1.400000E-05 RA223
5.843000E-01 1.400000E-05 RA223
5.940000E-01 1.400000E-05 RA223
5.987210E-01 9.500000E-04 RA223
6.093100E-01 5.700000E-04 RA223
6.191000E-01 3.500000E-05 RA223
6.235000E-01 8.000001E-05 RA223
6.317000E-01 4.000000E-06 RA223
6.417000E-01 1.700000E-05 RA223
6.461000E-01 4.000000E-06 RA223
6.969001E-01 7.000000E-06 RA223
7.113000E-01 3.600000E-05 RA223
7.184001E-01 1.400000E-05 RA223
7.284001E-01 2.800000E-06 RA223
7.328000E-01 2.800000E-06 RA223

129
1.230000E-02 2.700000E-01 FR223
2.027000E-02 1.600000E-02 FR223
2.960000E-02 2.400000E-04 FR223
2.978000E-02 5.400000E-04 FR223
3.169000E-02 1.500000E-05 FR223
4.350000E-02 2.400000E-05 FR223
4.400000E-02 1.500000E-05 FR223
4.980000E-02 2.800000E-02 FR223
5.009400E-02 3.400000E-01 FR223
6.143000E-02 3.900000E-05 FR223
6.231000E-02 1.800000E-04 FR223
7.350000E-02 1.500000E-05 FR223
7.965101E-02 8.700000E-02 FR223
8.543100E-02 1.630000E-02 FR223
8.847100E-02 2.700000E-02 FR223
8.908001E-02 6.000000E-04 FR223
9.388001E-02 6.600000E-04 FR223
9.943201E-02 3.200000E-03 FR223
1.001300E-01 6.200000E-03 FR223
1.024980E-01 2.300000E-03 FR223
1.110500E-01 5.400000E-05 FR223
1.345900E-01 5.299999E-03 FR223
1.555000E-01 3.000000E-05 FR223
1.733800E-01 1.200000E-05 FR223
1.846800E-01 2.500000E-03 FR223
2.007000E-01 3.000000E-05 FR223
2.049000E-01 9.800000E-03 FR223

2.056000E-01 6.600000E-05 FR223
2.106000E-01 1.080000E-04 FR223
2.188000E-01 9.600000E-05 FR223
2.347500E-01 3.000000E-02 FR223
2.360500E-01 3.000000E-04 FR223
2.456000E-01 2.900000E-04 FR223
2.502400E-01 2.030000E-04 FR223
2.546000E-01 6.300000E-05 FR223
2.561700E-01 3.000000E-04 FR223
2.629000E-01 3.900000E-05 FR223
2.728000E-01 4.500000E-05 FR223
2.807000E-01 1.200000E-05 FR223
2.860000E-01 1.000000E-04 FR223
2.896800E-01 3.000000E-04 FR223
2.965000E-01 1.500000E-05 FR223
2.999500E-01 2.100000E-04 FR223
3.044000E-01 1.500000E-04 FR223
3.077800E-01 1.480000E-04 FR223
3.126500E-01 1.700000E-04 FR223
3.146000E-01 2.400000E-05 FR223
3.192600E-01 4.900000E-03 FR223
3.298000E-01 2.700000E-04 FR223
3.343000E-01 1.100000E-04 FR223
3.395000E-01 6.700000E-04 FR223
3.425000E-01 2.000000E-04 FR223
3.505000E-01 3.000000E-05 FR223
3.693800E-01 9.900000E-04 FR223
4.344000E-01 2.400000E-05 FR223
4.396000E-01 3.300000E-06 FR223
4.445000E-01 1.200000E-05 FR223
4.529000E-01 1.800000E-05 FR223
4.575000E-01 8.999999E-06 FR223
4.693000E-01 2.400000E-05 FR223
4.754000E-01 6.300000E-05 FR223
4.809000E-01 1.500000E-05 FR223
4.934000E-01 2.700000E-05 FR223
5.069000E-01 2.400000E-05 FR223
5.167000E-01 3.600000E-05 FR223
5.248000E-01 4.800000E-05 FR223
5.331000E-01 2.100000E-05 FR223
5.372000E-01 5.700000E-05 FR223
5.398000E-01 6.600000E-05 FR223
5.454000E-01 3.300000E-06 FR223
5.523000E-01 3.000000E-05 FR223
5.563000E-01 1.200000E-05 FR223
5.689400E-01 5.500000E-04 FR223
5.761000E-01 1.200000E-05 FR223
5.813000E-01 1.500000E-05 FR223
5.923000E-01 3.600000E-05 FR223
5.969000E-01 8.999999E-06 FR223
6.007000E-01 6.000000E-06 FR223
6.076000E-01 2.400000E-05 FR223
6.136000E-01 1.200000E-05 FR223
6.327000E-01 2.400000E-05 FR223
6.637000E-01 1.200000E-05 FR223
6.719000E-01 6.000000E-06 FR223
7.083000E-01 1.500000E-05 FR223
7.231000E-01 4.400000E-04 FR223
7.241501E-01 1.600000E-04 FR223
7.374001E-01 1.000000E-05 FR223
7.424001E-01 1.200000E-05 FR223
7.463000E-01 2.200000E-04 FR223
7.536501E-01 1.000000E-04 FR223

```
7.572001E-01 1.000000E-04 FR223        4.370000E-01 9.500000E-06 FR221
7.626000E-01 2.700000E-05 FR223        4.463000E-01 1.700000E-05 FR221
7.658000E-01 2.400000E-04 FR223        4.683000E-01 1.400000E-05 FR221
7.758300E-01 4.600000E-03 FR223        5.378000E-01 5.100000E-05 FR221
7.808000E-01 3.300000E-05 FR223        5.623000E-01 5.100000E-05 FR221
7.849301E-01 1.500000E-04 FR223        5.685000E-01 1.200000E-05 FR221
7.876000E-01 3.150000E-05 FR223        5.769001E-01 3.000000E-05 FR221
7.922000E-01 6.000000E-06 FR223        6.520000E-01 4.100000E-06 FR221
7.962200E-01 1.000000E-04 FR223        6.650000E-01 9.200000E-06 FR221
8.037701E-01 5.900000E-04 FR223        8.093000E-01 9.200000E-06 FR221
8.060001E-01 1.500000E-05 FR223        8.919001E-01 3.400000E-06 FR221
8.124000E-01 2.100000E-04 FR223
8.165001E-01 1.500000E-05 FR223          1
8.232000E-01 7.800000E-05 FR223        5.497300E-01 1.140000E-03 RN220
8.259500E-01 5.400000E-04 FR223
8.339000E-01 1.500000E-05 FR223          34
8.362001E-01 8.000001E-05 FR223        1.110000E-02 9.600000E-02 RN219
8.421900E-01 5.400000E-05 FR223        7.686301E-02 5.000000E-02 RN219
8.468601E-01 5.400000E-04 FR223        7.929000E-02 8.399999E-02 RN219
8.636000E-01 3.600000E-05 FR223        8.925600E-02 1.010000E-02 RN219
8.674001E-01 1.800000E-05 FR223        8.980700E-02 1.900000E-02 RN219
8.765001E-01 4.100000E-04 FR223        9.231701E-02 7.200000E-03 RN219
8.781000E-01 3.600000E-05 FR223        1.306000E-01 1.300000E-03 RN219
8.931000E-01 2.800000E-05 FR223        2.240000E-01 1.400000E-05 RN219
8.967100E-01 1.500000E-04 FR223        2.712300E-01 1.080000E-01 RN219
9.076100E-01 1.400000E-04 FR223        2.935600E-01 7.300000E-04 RN219
9.113000E-01 8.999999E-06 FR223        3.218000E-01 9.000000E-07 RN219
9.136000E-01 4.500000E-06 FR223        3.308000E-01 9.699999E-06 RN219
9.265000E-01 1.800000E-05 FR223        3.735000E-01 2.500000E-06 RN219
9.412001E-01 3.300000E-05 FR223        3.831000E-01 4.300000E-06 RN219
9.493001E-01 3.600000E-06 FR223        4.018100E-01 6.600000E-02 RN219
9.580001E-01 3.900000E-06 FR223        4.054000E-01 2.500000E-06 RN219
9.692001E-01 3.600000E-06 FR223        4.369000E-01 3.000000E-06 RN219
9.752001E-01 1.800000E-06 FR223        4.616000E-01 1.600000E-06 RN219
9.787000E-01 7.500000E-06 FR223        4.893000E-01 6.300000E-06 RN219
9.894001E-01 1.500000E-06 FR223        5.176000E-01 4.400000E-04 RN219
9.943000E-01 1.200000E-06 FR223        5.561000E-01 5.000000E-07 RN219
9.993001E-01 2.100000E-06 FR223        5.641000E-01 1.500000E-05 RN219
1.025100E+00 1.500000E-06 FR223        5.766000E-01 9.000000E-07 RN219
                                       6.083000E-01 4.300000E-05 RN219
                                       6.199000E-01 3.200000E-06 RN219
    1                                  6.719000E-01 2.200000E-06 RN219
5.100001E-01 7.600000E-04 RN222        6.766600E-01 1.730000E-04 RN219
                                       7.081000E-01 3.200000E-06 RN219
    30                                 7.328000E-01 6.000000E-07 RN219
1.140000E-02 2.000000E-02 FR221        8.025000E-01 3.200000E-06 RN219
5.381000E-02 1.300000E-04 FR221        8.353000E-01 1.600000E-05 RN219
7.894800E-02 8.300000E-03 FR221        8.772001E-01 3.200000E-06 RN219
8.151700E-02 1.370000E-02 FR221        8.911000E-01 7.600000E-06 RN219
9.173001E-02 1.660000E-03 FR221        1.073700E+00 3.200000E-06 RN219
9.231500E-02 3.200000E-03 FR221
9.490000E-02 1.180000E-03 FR221
9.630001E-02 7.000000E-05 FR221          10
1.002500E-01 1.000000E-03 FR221        1.080000E-02 5.000000E-05 AT217
1.178200E-01 2.200000E-04 FR221        7.481501E-02 3.450000E-05 AT217
1.502100E-01 4.780000E-04 FR221        7.710700E-02 5.800000E-05 AT217
1.718300E-01 7.700000E-04 FR221        8.683001E-02 7.000000E-06 AT217
2.083000E-01 5.100000E-05 FR221        8.734901E-02 1.330000E-05 AT217
2.181200E-01 1.140000E-01 FR221        8.978400E-02 4.890000E-06 AT217
2.821200E-01 7.100000E-05 FR221        2.579000E-01 2.870000E-04 AT217
3.241000E-01 1.930000E-04 FR221        3.353300E-01 6.200000E-05 AT217
3.598600E-01 4.070000E-04 FR221        5.931000E-01 1.150000E-04 AT217
3.823400E-01 3.360000E-04 FR221        7.589000E-01 4.900000E-05 AT217
4.106400E-01 1.200000E-03 FR221
```

1
8.049001E-01 1.900000E-05 PO216

2
2.980000E-01 5.200000E-07 PO214
7.997000E-01 1.040000E-04 PO214

238
1.110000E-02 6.000000E-03 BI214B
7.686301E-02 3.970000E-03 BI214B
7.929000E-02 6.610000E-03 BI214B
8.925600E-02 8.000000E-04 BI214B
8.980700E-02 1.530000E-03 BI214B
9.231701E-02 5.660000E-04 BI214B
2.210000E-01 3.000000E-05 BI214B
2.300000E-01 4.000000E-05 BI214B
2.528000E-01 3.000000E-05 BI214B
2.688000E-01 2.000000E-04 BI214B
2.738000E-01 1.500000E-03 BI214B
2.809500E-01 6.000000E-04 BI214B
3.042000E-01 4.200000E-04 BI214B
3.333100E-01 8.000000E-04 BI214B
3.347800E-01 3.400000E-04 BI214B
3.489200E-01 1.200000E-03 BI214B
3.519000E-01 7.000000E-04 BI214B
3.560000E-01 7.000000E-05 BI214B
3.867700E-01 3.100000E-03 BI214B
3.888800E-01 3.700000E-03 BI214B
3.940500E-01 1.480000E-04 BI214B
3.960100E-01 2.900000E-04 BI214B
4.057400E-01 1.700000E-03 BI214B
4.280000E-01 1.100000E-05 BI214B
4.529200E-01 3.100000E-04 BI214B
4.547700E-01 3.000000E-03 BI214B
4.610000E-01 5.300000E-04 BI214B
4.697600E-01 1.290000E-03 BI214B
4.744100E-01 1.100000E-03 BI214B
4.859200E-01 2.200000E-04 BI214B
4.867000E-01 3.000000E-05 BI214B
4.879500E-01 2.800000E-04 BI214B
4.942000E-01 1.200000E-04 BI214B
4.969000E-01 6.899999E-05 BI214B
5.019600E-01 1.800000E-04 BI214B
5.199000E-01 1.600000E-04 BI214B
5.246000E-01 1.700000E-04 BI214B
5.280000E-01 4.000000E-05 BI214B
5.367700E-01 6.800000E-04 BI214B
5.430000E-01 8.400000E-04 BI214B
5.476000E-01 2.000000E-05 BI214B
5.727600E-01 7.400000E-04 BI214B
5.952300E-01 1.700000E-04 BI214B
6.000000E-01 8.000001E-05 BI214B
6.093121E-01 4.610000E-01 BI214B
6.157300E-01 6.000000E-04 BI214B
6.170000E-01 3.400000E-04 BI214B
6.264001E-01 5.000000E-05 BI214B
6.307900E-01 1.800000E-04 BI214B
6.331400E-01 5.500000E-04 BI214B
6.347200E-01 6.500000E-05 BI214B
6.396700E-01 3.000000E-04 BI214B
6.491800E-01 6.000000E-04 BI214B
6.515000E-01 1.000000E-05 BI214B
6.587000E-01 1.500000E-04 BI214B
6.611000E-01 4.700000E-04 BI214B
6.654530E-01 1.460000E-02 BI214B
6.774100E-01 6.000000E-05 BI214B
6.832200E-01 8.100000E-04 BI214B
6.876000E-01 6.899999E-05 BI214B
6.933000E-01 6.000000E-05 BI214B
6.979001E-01 5.100000E-04 BI214B
6.998200E-01 1.600000E-04 BI214B
7.031100E-01 4.720000E-03 BI214B
7.049001E-01 4.700000E-04 BI214B
7.088000E-01 1.700000E-04 BI214B
7.106700E-01 7.500000E-04 BI214B
7.198600E-01 3.790000E-03 BI214B
7.229800E-01 3.500000E-04 BI214B
7.338000E-01 4.300000E-04 BI214B
7.407300E-01 4.000000E-04 BI214B
7.528400E-01 1.300000E-03 BI214B
7.683560E-01 4.940000E-02 BI214B
7.697001E-01 3.000000E-04 BI214B
7.861000E-01 3.100000E-03 BI214B
7.886000E-01 1.500000E-04 BI214B
8.061740E-01 1.220000E-02 BI214B
8.150001E-01 3.800000E-04 BI214B
8.211800E-01 1.580000E-03 BI214B
8.263000E-01 1.100000E-03 BI214B
8.323901E-01 2.800000E-04 BI214B
8.404000E-01 8.999999E-05 BI214B
8.471600E-01 2.600000E-04 BI214B
8.730701E-01 1.800000E-04 BI214B
8.780301E-01 1.200000E-04 BI214B
9.042900E-01 8.500000E-04 BI214B
9.157400E-01 2.600000E-04 BI214B
9.178000E-01 5.000000E-05 BI214B
9.302000E-01 3.300000E-04 BI214B
9.340610E-01 3.030000E-02 BI214B
9.341000E-01 5.000000E-04 BI214B
9.345000E-01 1.000000E-04 BI214B
9.386501E-01 1.300000E-04 BI214B
9.396000E-01 1.800000E-04 BI214B
9.433401E-01 1.700000E-04 BI214B
9.498000E-01 5.500000E-05 BI214B
9.522001E-01 6.000000E-05 BI214B
9.616100E-01 1.200000E-04 BI214B
9.640800E-01 3.620000E-03 BI214B
9.650000E-01 1.000000E-04 BI214B
9.761800E-01 1.900000E-04 BI214B
9.914901E-01 1.000000E-04 BI214B
1.013800E+00 8.300000E-05 BI214B
1.021000E+00 1.400000E-04 BI214B
1.032370E+00 7.800000E-04 BI214B
1.033300E+00 2.400000E-04 BI214B
1.038000E+00 8.300000E-05 BI214B
1.045600E+00 2.600000E-04 BI214B
1.051960E+00 3.150000E-03 BI214B
1.067200E+00 2.700000E-04 BI214B
1.069960E+00 2.750000E-03 BI214B
1.103640E+00 1.000000E-03 BI214B
1.104790E+00 7.700000E-04 BI214B
1.118900E+00 4.000000E-04 BI214B
1.120287E+00 1.510000E-01 BI214B
1.130290E+00 4.000000E-04 BI214B
1.133660E+00 2.480000E-03 BI214B
1.155190E+00 1.630000E-02 BI214B

1.155600E+00	1.600000E-04	BI214B		2.147900E+00	1.400000E-04	BI214B
1.156000E+00	7.000000E-05	BI214B		2.160400E+00	1.800000E-05	BI214B
1.167300E+00	1.200000E-04	BI214B		2.176500E+00	3.200000E-05	BI214B
1.172980E+00	5.100000E-04	BI214B		2.192580E+00	3.400000E-04	BI214B
1.207680E+00	4.510000E-03	BI214B		2.204210E+00	5.080000E-02	BI214B
1.226700E+00	1.800000E-04	BI214B		2.251600E+00	5.500000E-05	BI214B
1.230600E+00	1.500000E-04	BI214B		2.260300E+00	8.700000E-05	BI214B
1.238110E+00	5.790000E-02	BI214B		2.266510E+00	1.800000E-04	BI214B
1.279000E+00	1.200000E-04	BI214B		2.270900E+00	1.300000E-05	BI214B
1.280960E+00	1.430000E-02	BI214B		2.284300E+00	5.100000E-05	BI214B
1.284000E+00	1.100000E-04	BI214B		2.287650E+00	4.600000E-05	BI214B
1.285100E+00	1.700000E-04	BI214B		2.293400E+00	3.050000E-03	BI214B
1.303760E+00	1.120000E-03	BI214B		2.310200E+00	1.400000E-05	BI214B
1.316960E+00	8.000000E-04	BI214B		2.312400E+00	8.999999E-05	BI214B
1.330000E+00	1.100000E-04	BI214B		2.319300E+00	4.000000E-06	BI214B
1.341490E+00	2.200000E-04	BI214B		2.325000E+00	1.700000E-05	BI214B
1.377669E+00	4.000000E-02	BI214B		2.331300E+00	2.210000E-04	BI214B
1.385310E+00	7.570000E-03	BI214B		2.348000E+00	1.400000E-06	BI214B
1.392500E+00	1.900000E-04	BI214B		2.353500E+00	4.000000E-06	BI214B
1.401500E+00	1.270000E-02	BI214B		2.361000E+00	1.700000E-05	BI214B
1.407980E+00	2.150000E-02	BI214B		2.369000E+00	2.700000E-05	BI214B
1.415800E+00	4.810000E-03	BI214B		2.376900E+00	8.800000E-05	BI214B
1.419700E+00	5.100000E-05	BI214B		2.390800E+00	1.600000E-05	BI214B
1.470900E+00	9.200000E-05	BI214B		2.405100E+00	4.100000E-06	BI214B
1.479150E+00	5.100000E-04	BI214B		2.423270E+00	4.600000E-05	BI214B
1.509228E+00	2.110000E-02	BI214B		2.444700E+00	8.000001E-05	BI214B
1.515500E+00	6.899999E-05	BI214B		2.447860E+00	1.570000E-02	BI214B
1.538500E+00	3.760000E-03	BI214B		2.482800E+00	1.500000E-05	BI214B
1.543320E+00	2.000000E-03	BI214B		2.505400E+00	5.700000E-05	BI214B
1.583220E+00	6.900000E-03	BI214B		2.550700E+00	4.600000E-06	BI214B
1.594730E+00	2.500000E-03	BI214B		2.553000E+00	1.000000E-06	BI214B
1.595000E+00	5.000000E-05	BI214B		2.562000E+00	1.800000E-06	BI214B
1.598000E+00	6.000000E-05	BI214B		2.564000E+00	1.400000E-06	BI214B
1.599310E+00	2.300000E-03	BI214B		2.604500E+00	4.000000E-06	BI214B
1.636300E+00	1.200000E-04	BI214B		2.630900E+00	8.000000E-06	BI214B
1.637000E+00	6.000000E-05	BI214B		2.662400E+00	3.000000E-06	BI214B
1.657000E+00	4.600000E-04	BI214B		2.694700E+00	3.100000E-04	BI214B
1.661280E+00	1.150000E-02	BI214B		2.699400E+00	2.800000E-05	BI214B
1.665800E+00	8.300000E-05	BI214B		2.719300E+00	1.800000E-05	BI214B
1.683990E+00	2.160000E-03	BI214B		2.769900E+00	2.500000E-04	BI214B
1.711000E+00	1.800000E-05	BI214B		2.785900E+00	5.500000E-05	BI214B
1.729595E+00	2.920000E-02	BI214B		2.827000E+00	2.300000E-05	BI214B
1.751400E+00	8.999999E-06	BI214B		2.861100E+00	3.800000E-06	BI214B
1.764494E+00	1.540000E-01	BI214B		2.880300E+00	9.200000E-05	BI214B
1.813730E+00	1.100000E-04	BI214B		2.893500E+00	6.000000E-05	BI214B
1.819200E+00	7.000000E-06	BI214B		2.921900E+00	1.400000E-04	BI214B
1.838360E+00	3.600000E-03	BI214B		2.928600E+00	1.100000E-05	BI214B
1.847420E+00	2.110000E-02	BI214B		2.934600E+00	4.600000E-06	BI214B
1.873160E+00	2.190000E-03	BI214B		2.978900E+00	1.380000E-04	BI214B
1.890300E+00	8.000000E-04	BI214B		3.000000E+00	8.800000E-05	BI214B
1.895920E+00	1.600000E-03	BI214B		3.053900E+00	2.100000E-04	BI214B
1.898700E+00	5.700000E-04	BI214B		3.081700E+00	4.800000E-05	BI214B
1.935500E+00	4.100000E-04	BI214B		3.094000E+00	4.400000E-06	BI214B
1.994600E+00	5.000000E-05	BI214B		3.142600E+00	1.200000E-05	BI214B
2.010780E+00	4.700000E-04	BI214B		3.149000E+00	8.800000E-07	BI214B
2.016700E+00	6.000000E-05	BI214B		3.160600E+00	3.200000E-06	BI214B
2.021600E+00	2.000000E-04	BI214B		3.183600E+00	1.290000E-05	BI214B
2.052940E+00	6.900000E-04	BI214B				
2.085100E+00	9.100000E-05	BI214B		30		
2.089700E+00	5.000000E-04	BI214B		1.080000E-02	1.330000E-01	PB214
2.109920E+00	8.799999E-05	BI214B		5.322750E-02	1.200000E-02	PB214
2.118550E+00	1.140000E-02	BI214B		7.481501E-02	6.330000E-02	PB214
2.120000E+00	7.000000E-05	BI214B		7.710700E-02	1.050000E-01	PB214

8.683001E-02 1.280000E-02 PB214
8.734901E-02 2.440000E-02 PB214
8.978400E-02 9.000000E-03 PB214
1.181600E-01 9.399999E-04 PB214
1.962000E-01 6.900000E-04 PB214
2.056800E-01 1.150000E-04 PB214
2.384000E-01 8.000001E-05 PB214
2.419970E-01 7.430000E-02 PB214
2.588700E-01 5.240000E-03 PB214
2.748000E-01 4.740000E-03 PB214
2.952240E-01 1.930000E-01 PB214
2.987600E-01 1.000000E-04 PB214
3.052600E-01 3.100000E-04 PB214
3.143200E-01 7.800000E-04 PB214
3.238300E-01 2.800000E-04 PB214
3.519320E-01 3.760000E-01 PB214
4.620000E-01 2.210000E-03 PB214
4.804300E-01 3.200000E-03 PB214
4.870900E-01 4.220000E-03 PB214
5.110000E-01 3.200000E-04 PB214
5.336600E-01 1.860000E-03 PB214
5.384100E-01 2.000000E-04 PB214
5.438100E-01 6.900000E-04 PB214
5.801300E-01 3.520000E-03 PB214
7.859601E-01 1.070000E-02 PB214
8.390400E-01 5.870000E-03 PB214

1

7.788000E-01 4.800000E-05 PO213

3

3.238100E-01 1.650000E-03 BI213A
5.449001E-01 1.650000E-04 BI213A
8.680000E-01 1.270000E-04 BI213A

24

1.110000E-02 1.540000E-02 BI213B
7.686301E-02 1.080000E-02 BI213B
7.929000E-02 1.790000E-02 BI213B
8.925600E-02 2.170000E-03 BI213B
8.980700E-02 4.160000E-03 BI213B
9.231701E-02 1.540000E-03 BI213B
1.476600E-01 1.500000E-04 BI213B
2.928000E-01 4.290000E-03 BI213B
4.028000E-01 1.000000E-06 BI213B
4.404500E-01 2.594000E-01 BI213B
5.749000E-01 2.500000E-05 BI213B
6.009001E-01 4.200000E-05 BI213B
6.049401E-01 2.300000E-05 BI213B
6.597500E-01 3.610000E-04 BI213B
7.108100E-01 1.110000E-04 BI213B
8.073601E-01 2.920000E-03 BI213B
8.265201E-01 7.100000E-05 BI213B
8.679800E-01 1.110000E-04 BI213B
8.866600E-01 1.020000E-05 BI213B
1.003550E+00 5.300000E-04 BI213B
1.045700E+00 1.800000E-04 BI213B
1.100130E+00 2.590000E-03 BI213B
1.119290E+00 5.200000E-04 BI213B
1.328200E+00 4.000000E-06 BI213B

15

1.030000E-02 7.000000E-02 BI212A
3.985700E-02 1.060000E-02 BI212A
7.083201E-02 5.400000E-04 BI212A
7.287300E-02 9.050000E-04 BI212A
8.211500E-02 1.090000E-04 BI212A
8.257400E-02 2.090000E-04 BI212A
8.486500E-02 7.620000E-05 BI212A
2.882000E-01 3.370000E-03 BI212A
3.280300E-01 1.250000E-03 BI212A
4.337000E-01 1.700000E-04 BI212A
4.529800E-01 3.630000E-03 BI212A
4.730000E-01 5.000000E-04 BI212A
4.933000E-01 1.800000E-05 BI212A
5.760000E-01 1.800000E-06 BI212A
6.200000E-01 5.000000E-06 BI212A

18

1.110000E-02 5.750000E-04 BI212B
7.686301E-02 3.930000E-04 BI212B
7.929000E-02 6.550000E-04 BI212B
8.925600E-02 7.910000E-05 BI212B
8.980700E-02 1.520000E-04 BI212B
9.231701E-02 5.610000E-05 BI212B
1.802000E-01 3.200000E-05 BI212B
7.273300E-01 6.670000E-02 BI212B
7.853701E-01 1.102000E-02 BI212B
8.934081E-01 3.780000E-03 BI212B
9.521201E-01 1.700000E-03 BI212B
1.073600E+00 1.600000E-04 BI212B
1.078620E+00 5.640000E-03 BI212B
1.512700E+00 2.900000E-03 BI212B
1.620500E+00 1.470000E-02 BI212B
1.679700E+00 5.800000E-04 BI212B
1.800200E+00 4.100000E-05 BI212B
1.806000E+00 9.000000E-04 BI212B

11

1.080000E-02 1.430000E-01 PB212
7.481501E-02 1.028000E-01 PB212
7.710700E-02 1.710000E-01 PB212
8.683001E-02 2.070000E-02 PB212
8.734901E-02 3.970000E-02 PB212
8.978400E-02 1.460000E-02 PB212
1.151830E-01 5.960000E-03 PB212
1.766800E-01 5.200000E-04 PB212
2.386320E-01 4.360000E-01 PB212
3.000870E-01 3.300000E-02 PB212
4.152000E-01 1.310000E-04 PB212

9

1.060000E-02 8.100000E-05 PO211
7.280500E-02 5.780000E-05 PO211
7.496901E-02 9.600000E-05 PO211
8.445000E-02 1.170000E-05 PO211
8.493800E-02 2.230000E-05 PO211
8.730001E-02 8.200000E-06 PO211
3.282000E-01 3.300000E-05 PO211
5.696501E-01 5.450000E-03 PO211
8.978000E-01 5.610000E-03 PO211

7

1.030000E-02 9.600000E-03 BI211A
7.083201E-02 7.340000E-03 BI211A
7.287300E-02 1.230000E-02 BI211A
8.211500E-02 1.490000E-03 BI211A
8.257400E-02 2.850000E-03 BI211A
8.486500E-02 1.038000E-03 BI211A
3.510600E-01 1.292000E-01 BI211A

29

1.080000E-02 5.100000E-03 PB211
6.542000E-02 7.700000E-04 PB211
7.481501E-02 2.260000E-03 PB211
7.710700E-02 3.760000E-03 PB211
8.683001E-02 4.550000E-04 PB211
8.734901E-02 8.699999E-04 PB211
8.978400E-02 3.200000E-04 PB211
9.500001E-02 1.800000E-04 PB211
3.135900E-01 3.100000E-04 PB211
3.429100E-01 3.500000E-04 PB211
3.620720E-01 4.300000E-04 PB211
4.048530E-01 3.780000E-02 PB211
4.270880E-01 1.760000E-02 PB211
4.300000E-01 6.000000E-05 PB211
5.041200E-01 5.800000E-05 PB211
6.093800E-01 4.300000E-04 PB211
6.766900E-01 1.300000E-04 PB211
7.046400E-01 4.620000E-03 PB211
7.665101E-01 6.169999E-03 PB211
8.320100E-01 3.520000E-02 PB211
8.659300E-01 5.900000E-05 PB211
9.510000E-01 2.200000E-04 PB211
1.014640E+00 1.730000E-04 PB211
1.080160E+00 1.230000E-04 PB211
1.103520E+00 4.600000E-05 PB211
1.109480E+00 1.150000E-03 PB211
1.196330E+00 1.020000E-04 PB211
1.234300E+00 1.300000E-05 PB211
1.270710E+00 6.800000E-05 PB211

7

1.060000E-02 3.820000E-08 PO210
7.280500E-02 2.710000E-08 PO210
7.496901E-02 4.520000E-08 PO210
8.445000E-02 5.460000E-09 PO210
8.493800E-02 1.050000E-08 PO210
8.730001E-02 3.830000E-09 PO210
8.031000E-01 1.210000E-05 PO210

2

1.080000E-02 2.360000E-01 PB210
4.653900E-02 4.250000E-02 PB210

22

1.060000E-02 1.900000E-01 TL210
7.280500E-02 2.300000E-02 TL210
7.496901E-02 3.800000E-02 TL210
8.300000E-02 2.000000E-02 TL210
8.445000E-02 4.600000E-03 TL210
8.493800E-02 8.800000E-03 TL210
8.730001E-02 3.200000E-03 TL210
9.700000E-02 4.000000E-02 TL210
2.960000E-01 7.900000E-01 TL210

4.800000E-01 2.000000E-02 TL210
7.996000E-01 9.895999E-01 TL210
8.600000E-01 6.900000E-02 TL210
1.070000E+00 1.200000E-01 TL210
1.110000E+00 6.900000E-02 TL210
1.210000E+00 1.700000E-01 TL210
1.316000E+00 2.100000E-01 TL210
1.410000E+00 4.900000E-02 TL210
1.590000E+00 2.000000E-02 TL210
2.010000E+00 6.900000E-02 TL210
2.270000E+00 3.000000E-02 TL210
2.360000E+00 8.000000E-02 TL210
2.430000E+00 9.000000E-02 TL210

11

1.060000E-02 8.700000E-02 TL209
7.280500E-02 6.210000E-02 TL209
7.496901E-02 1.040000E-01 TL209
8.445000E-02 1.250000E-02 TL209
8.493800E-02 2.400000E-02 TL209
8.730001E-02 8.800000E-03 TL209
1.172110E-01 8.430000E-01 TL209
4.651300E-01 9.690000E-01 TL209
9.201301E-01 6.100000E-03 TL209
1.239160E+00 4.600000E-03 TL209
1.567090E+00 9.980000E-01 TL209

37

1.060000E-02 2.750000E-02 TL208
7.280500E-02 2.010000E-02 TL208
7.496901E-02 3.350000E-02 TL208
8.445000E-02 4.040000E-03 TL208
8.493800E-02 7.760000E-03 TL208
8.730001E-02 2.830000E-03 TL208
2.114000E-01 1.800000E-03 TL208
2.333600E-01 3.100000E-03 TL208
2.526100E-01 7.799999E-03 TL208
2.773710E-01 6.600000E-02 TL208
4.859500E-01 4.899999E-04 TL208
5.107700E-01 2.260000E-01 TL208
5.831870E-01 8.500000E-01 TL208
5.877001E-01 6.000000E-04 TL208
6.501000E-01 5.000000E-04 TL208
7.052000E-01 2.200000E-04 TL208
7.220400E-01 2.400000E-03 TL208
7.487000E-01 4.600000E-04 TL208
7.631301E-01 1.790000E-02 TL208
8.212001E-01 4.100000E-04 TL208
8.605570E-01 1.250000E-01 TL208
8.833000E-01 3.100000E-04 TL208
9.276000E-01 1.250000E-03 TL208
9.827000E-01 2.050000E-03 TL208
1.093900E+00 4.300000E-03 TL208
1.125700E+00 5.000000E-05 TL208
1.160800E+00 1.100000E-04 TL208
1.185200E+00 1.700000E-04 TL208
1.282800E+00 5.200000E-04 TL208
1.381100E+00 7.000000E-05 TL208
1.647500E+00 2.000000E-05 TL208
1.744000E+00 2.000000E-05 TL208
2.614511E+00 9.975399E-01 TL208
3.197700E+00 4.000000E-05 TL208
3.475100E+00 1.500000E-05 TL208

```
3.708400E+00  2.000000E-05  TL208
3.960900E+00  1.500000E-05  TL208

  9
1.060000E-02  2.090000E-05  TL207
7.280500E-02  1.570000E-05  TL207
7.496901E-02  2.610000E-05  TL207
8.445000E-02  3.150000E-06  TL207
8.493800E-02  6.050000E-06  TL207
8.730001E-02  2.210000E-06  TL207
3.281000E-01  1.400000E-05  TL207
5.696200E-01  1.590000E-05  TL207
8.977700E-01  2.600000E-03  TL207
```